AAPG REPRINT SERIES NO. 15

Carbonate Rocks III: Organic Reefs

Selected Papers Reprinted from

AAPG BULLETIN

Published by The American Association of Petroleum Geologists
Tulsa, Oklahoma, U.S.A., 1975

Carbonate Rocks III: Organic Reefs: Preface

Organic reefs and reef-related facies of several ages have proved to be highly productive petroleum reservoirs in many parts of the world. They have been principal exploration objectives from the deserts of the Middle East, to the tropical coasts of Mexico, to the subarctic plains of Western Canada. The discovery and exploitation of reservoirs in reef and reef-related facies require, perhaps more than any other sort of petroleum occurrence, a dependence upon broadly based geologic investigations. It is in the discovery and successful development of reef or reef-related reservoirs that the interdependent roles of the regional stratigrapher, sedimentologist, paleontologist, carbonate petrographer, structural geologist, geochemist, and geophysicist have been brought to bear most effectively.

The prominence of papers dealing with reef problems in the AAPG *Bulletin*, Special Publications, and Memoirs is evidence of the importance of reefs to the petroleum geologist. Reefs have been the subject of over 75 papers occupying almost 1,500 pages of text in the *Bulletin* alone during the 59-year history of the Association. Several Special Publications and each of at least eight Memoirs contain one or more reef papers. In addition, over 90 papers on reefs have been presented at Association meetings. In all, the Association has published more than 2,200 pages on reefs in over 90 major papers and over 95 abstracts, notes, and book reviews. More than half of the major reef papers have been selected for republication in various volumes of the Reprint Series.

The present volume contains 11 papers representing many of the most important modern contributions to reef geology and illustrating the application of sedimentologic, stratigraphic, paleontologic, geochemical, petrographic, and structural principles to the investigation and interpretation of ancient reefs and their relationships to the origin and accumulation of oil and gas. Included are Twenhofel's, Henson's, and Link's papers that have become classic references for students of ancient reefs. Papers by Cloud and Stanton summarize and synthesize facies data by which regional relationships of reefs have been ascertained and by which hydrocarbon reservoirs within reefs can be developed. The paper by Chester illustrates a geochemical method for defining reef facies by the distribution of trace elements. The papers by Peterson and Bergenback and Terriere describe and interpret reef facies entirely on the basis of subsurface data. In his paper, Achauer challenges the traditional interpretation of the Permian Capitan Formation as a barrier-reef complex and presents data to support his view that the Capitan originated as a linear organic bank generally devoid of frame-building organisms and reef-derived detritus. Dunham and Braithwaite suggest in their papers refinements of reef terminology and application of criteria for recognition of reefs and their separation from other carbonate-rock bodies in the geologic column.

This array of papers presents concepts, models, and techniques of fundamental value in helping both the novice and experienced geologist in recognizing and interpreting ancient reefs and in developing oil and gas deposits in reef facies.

The papers in this volume have been selected by Robert H. Dott, Sr., and Peggy Rice, of the AAPG staff, John Shelton, Editor of the AAPG, and Jules Braunstein, Chairman of the Committee for Publication. They were assembled by E. M. Tidwell, of the AAPG staff.

BOB F. PERKINS
The University of Texas at Arlington
Arlington, Texas
September 14, 1975

Reprinted from
BULLETIN OF THE AMERICAN ASSOCIATION OF PETROLEUM GEOLOGISTS
VOL. 34, NO. 2 (FEBRUARY, 1950), PP. 182-202, 2 FIGS.

CORAL AND OTHER ORGANIC REEFS IN GEOLOGIC COLUMN[1]

W. H. TWENHOFEL[2]
Orlando, Florida

ABSTRACT

The several definitions and meanings of the word reef are considered and also the application of the term bioherm proposed by Cumings and Shrock. Reefs from the geological point of view are defined and the characteristics of such reefs are presented. Reefs in the geologic column from the Huronian to and including the Tertiary are described.

INTRODUCTION

When the word reef is mentioned, the ordinary layman thinks of tropical islands, Robinson Crusoe, Swiss Family Robinson, palms loaded with cocoanuts, dusky maidens in grass skirts or no skirts at all, and, if he listens to the radio, perhaps of Peter Paul's Mounds and Almond Joy; the seaman thinks of rocks just beneath or awash with the surface of the sea; the Australian miner of steeply inclined rocks; but, if the hearer has had some training in geology, the thought may be of the Darwin-Dana subsidence hypothesis of the formation of barrier reefs and atolls, perhaps the Murray-Agassiz solution theory, the Daly glacial-control hypothesis, or possibly the antecedent-platform hypothesis of Hoffmeister and Ladd and their predecessors who have advanced that hypothesis.

The various points of view emphasize the different meanings attached to the word. And these are not all. On the Island of Anticosti and the adjacent Mingan Islands the wave-cut terraces which surround these islands are known as reefs. *Webster's Dictionary* defines reef as

> A range or ridge of rocks lying at or near the surface of the water, especially one of coral; A lode, vein, or ledge (Australia); A shoal or bank in the sea (use by fishermen); A variety of sponge (a trade name); The part of a sail that is wrapped up and secured by points, usually to a yard or boom, when it becomes necessary to shorten or diminish the size of a sail.

The word reef as used by biologists and geologists has been more or less differently defined, but all have understood that it referred to structures in the building of which corals played a considerable part. Vaughan (1911) gives the following description of a coral reef.

> A coral reef is a ridge or mound of limestone, the upper surface of which lies, or lay at the time of formation, near the level of the sea, and is predominantly composed of calcium carbonate secreted by organisms, of which the most important are corals.

It is now well known that most existing coral reefs do not have corals as the most important agents in their formation. Walther, in 1911, defined a coral reef as follows.

> A coral reef is an isolated calcareous mass, rising above the sea bottom, and essentially built by branching corals which collect detrital materials and prevent these from eventually being spread over the sea bottom (Cumings' translation, 1932).

[1] Read before the Association at St. Louis, March 16, 1949. Manuscript received, May 15, 1949.

[2] Professor of geology, retired, University of Wisconsin.

He recognized that the major part of the mass of a reef does not consist of coral colonies. Walther's definition may hold for some reefs, but few reefs in the geologic column known to the writer contain many branching corals. These play little part in the formation of the organic reefs of the Paleozoic in which corals played an important role.

The many meanings of the word reef led Cumings and Shrock (1928) and Cumings (1932) to suggest that the word be abandoned as applicable to the organic structures in the geologic column generally known as reefs and they proposed to substitute therefore the word bioherm from the Greek stem *bio* and the Greek word *herma*, the latter meaning a reef, bank, or mound. The term was intended to apply to reef-like, mound-like, lens-like or other circumscribed structures of organic origin which are embedded in and surrounded by sediments of other lithologic nature. The suggested term has merit and should be extensively used, but the word reef is so generally ingrained by use that complete displacement is unlikely.

The writer's understanding of an organic reef or bioherm is that it is an aggregate of organic materials of significant dimensions built in part by colonial organisms, mainly, but not necessarily marine, that lived near, but below, the surface of the water (they may have some exposure at low tide) and constructed exoskeletal matter that became more or less adherent to previously formed similar matter. Nooks and depressions between the structures made by the colonial organisms, which may have as much space as, or greater than, the structures made by them, formed excellent dwelling places for non-colonial organisms which sought these places for protection or for food. Organic reefs develop substantially large vertical dimensions as compared with the dimensions of adjacent contemporaneous deposits. Organisms responsible for the somewhat rigid framework of a reef are permanent dwellers on it and their exoskeletons or secretions remain to make a part of it after their deaths. As the structure developed by the building of the forming organisms has considerable rigidity, it does not change volume under the weight of any overburden placed upon it. In a sense, the complete structure is reinforced by the overlapping colonial growths and this reinforcement enables the accumulation to grow upward at much steeper angles than is possible for most sedimentary materials and certainly for all clastic materials. Inclinations beyond the vertical are possible. Contact of a reef with marginal sediments is irregular and at angles that are not possible with ordinary sediments. This contact may be horizontal both above and below, as shown in Figure 1, and thence it may range to vertical.

Reefs are places of intensive and extensive competitive organic activity. The organisms that build reefs are food for other organisms and where food is abundant, as it is likely to be on reefs, there are those to devour it and these may throng a reef. After the deaths of the reef-builders, their constructions may be riddled by boring organisms in their search for food or to construct cavities in which to live. In this way the framework of the structure may be greatly impaired and reduced in part to calcareous sand and mud, and this destruction may be so complete that the resulting materials may show little evidence of once having been a rigid framework. Diagenesis may also affect the materials of a reef. Permeating waters are generally warm and they are likely to be charged with carbon dioxide derived from decaying organic matter. These conditions favor solution of reef materials, possibly their recrystallization, and perhaps some changes in composition. The waters may also contain much matter in solution and some of this may be deposited and cement the reef materials.

FIG. 1.—Photograph of reef forming Hobergen Klint on south end of Gotland. Core of reef is 150 yards in diameter, and it is 75 feet thick. Irregular contact of reef with marginal rock shown by white line. Reef interfingers with marginal strata. Core is little altered, and splendidly preserved fossils are present in spaces within coralline framework. Photograph after Munthe.

If a reef is built on deposits of mud or silt of which the supporting competency is not great, these may flow from beneath the load and cause underlying sediments to dip toward and beneath the reef. This has been reported from the Silurian reefs of Gotland, by Cumings and Shrock (1928, p. 613) for some of the reefs of Indiana, and also from other localities.

Sediments deposited marginal to a reef are likely to be calcareous sands, perhaps gravels, passing outward into calcareous muds on the seaward side and limited calcareous sands and mostly muds on the landward side. The sediments adjacent to a reef on the seaward side form the forereef. These are largely derived from erosion of reef materials and partly from organisms that lived on the fore-reef or in the overlying waters. The forereef in some respects may be looked on as a kind of talus slope. The backreef in some ways is not unlike the forereef, but the composing materials are not ordinarily so coarse and do not incline so steeply from the reef. Sediments of all kinds and origins may fill cavities between the colonial constructions of a reef, but the contributions are largely calcareous and probably most are derived directly from erosion of reef materials and from organisms that dwell in the cavities.

The sediments of the forereef incline away from the reef, and for forereefs seen in the geologic column these angles of inclination may be very high and are commonly greater than the angles of repose of the materials concerned, so that some cementation probably takes place concurrently with deposition. Perhaps some steep angles are due to slumping after some cementation. Angles as great as 45° are not uncommon and some exceed 60°. Cumings and Shrock (1928) report 65° at a Silurian reef near Wabash, Indiana.

Lime muds deposited in connection with reefs are filled with water at time of deposition. As this water is expelled, compaction takes place and original dips may be increased. Ultimately reef-building comes to a close and sediments may be deposited over the reefs. These arch over in the forms of domes or anticlines and on the margins have dips approximately equal to the underlying sediments that drape around the reefs. Dips are steeper on the seaward than the landward side unless a reef has the form of a dome some distance from the land, in which case dips may be approximately the same on all sides. This is commonly the case in the reefs in the lower and middle Paleozoic. Additional deposition may produce additional compacting of bordering sediments, but little to no compacting of reef materials. The apparent structure may be thus intensified and it may be projected upward through many feet of sediments.

Materials of a reef have an original porosity made by spaces between tabulae of corals, between diaphragms of bryozoans, between laminae of algal growths, and within shells of other organisms. Permeability may be low. Some of the original porosity may be destroyed by boring organisms, but other porosity is formed at the same time and permeability may be increased. The drilling on Bikini showed that most of the material encountered in the wells was unconsolidated and much was in the form of calcareous sand. Perhaps some of the sand came from penetration of the forereef, but it may well have been produced by boring organisms. The drilling of Funafuti encountered hard dolomitic limestone from 748 feet to the bottom at 1,114 feet, but to the upper depth the materials were either poorly cemented or unconsolidated. The drilling on the Island of Kita-Daito-Zima (Endo) also found unconsolidated materials and the drilling on the Great Barrier Reef of Australia found that the materials to the entire depths of 600 and 732 feet were loosely coherent. Obviously, reefs retain much porosity and probably considerable permeability to depths of hundreds of feet. Some additional porosity and permeability may be created by leaching if permeating waters are not saturated with calcium carbonate, and the drilling of

Funafuti showed that leaching had taken place at some levels. Porosity and permeability may be decreased by deposition of materials in solution if permeating waters are saturated with substances that may be deposited. The material deposited is likely to be calcium carbonate, but it may be magnesium carbonate or silica. The extent of these different processes can not be forecast for fossil reefs.

If reef-building comes to an end by reason of submergence faster than the organisms can build upward, a reef becomes a high place on the bottom from which the finer sediments would probably be washed away and deposited over bordering low places. Only those sediments would remain on the high places which the bottom waters were unable to remove. Trask, Revelle, and Revelle and Shepard have shown that this is what takes place on the high places on the bottom off the coast of California. These relatively coarse sediments have original porosity and permeability of which both might be decreased or even eliminated by cementation. Thus a reef may have porosity in the reef itself, in directly overlying sediments, and in the bordering sands of the forereef and the backreef.

Dimensions of reefs may be very great. The Great Barrier Reef of Australia has a length of more than 1,000 miles and a thickness, as suggested by the two holes drilled into it, about 700 miles apart, of 375 and 506 feet. Few ancient reefs approach this dimension although Cumings and Shrock (1928) state that the reef area in the Silurian around the Michigan basin was a close rival. The drilling on Funafuti suggested a thickness of at least 637 feet and possibly more than 1,114 feet. The drilling on Bikini showed a thickness of 65–75 feet. That on the Island of Kita-Daito-Zima showed a thickness of 340 feet and those on the Island of Marotoea, northwest of Borneo, a thickness of 1,400 feet. No ancient reef made by coral as an important contributor is known to have anything like a comparable thickness. Cumings and Shrock report single reefs in the Silurian of Indiana to range to a mile in diameter and to 100 feet in thickness (1928, p. 611). Some reefs in the Permian to which algae seem to have been the most important contributors may be as thick as any reefs in recent seas.

Reef-building organisms can not build above high-tide level and, when that level is attained, upward building of a reef ceases. Lateral expansion may then begin, particularly in the seaward direction as that is generally the direction of better aerated and food-containing water. The organisms build outward over the forereef and on all sides on a circular reef isolated from the shore. A period of halting in a rise of sea-level should be indicated by a widening of the reef.

The rocks of reefs have little stratification excepting as such is produced by the plastering of organisms, one over another. Excellent limited stratification may be present in spaces between colonial growths. These receive sediments eroded from the reef or introduced from without.

As stated, boring organisms may completely destroy reef materials and structure and reduce the framework to sands and smaller particles so that the materials of a reef may become almost devoid of recognizable macroscopic organic matter. However, some fossil reefs show little sign of such destructive work and contain extremely well preserved organic remains. Some of the Silurian reefs of Gotland and Anticosti Island are filled with excellently preserved corals and little seems to have been done to them after the deaths of their builders. Colonies of other organisms between the colonial bodies in places are buried in fine-grained sediments deposited from suspension, and some of these are wonderfully well preserved. Some reefs in the Silurian of eastern Wisconsin seem almost barren of organic remains and Cumings and Shrock report this to be the case in Silurian reefs of Indiana. It may be that the reefs that contain few fossils are largely composed of algal constructions, and, as is well known, these commonly show little or no evidence of organic structure.

Some reefs are almost completely dolomitized as are those in the Silurian of eastern Wisconsin and of Indiana, whereas the reefs of Gotland and Anticosti Island show no dolomitization. The coral reef of Funafuti showed dolomitization from the depth of 748 feet to the bottom of the hole at 1,114 feet. On the other hand, a drill hole on Bikini to the depth of 2,556 feet showed no dolomitization. The well drilled on the Island of Kita-Baito-Zima, a small island south of Japan and east of Okinawa, to the depth of 1,416 feet showed dolomitization to the depth of 340 feet. The two holes drilled in the Great Barrier Reef to the depths of 600 and 732 feet showed no dolomitization. Why the dolomitization at some depths in one hole and not in another at the same level remains to be learned.

Organisms responsible for reef formation are many and to some degree most benthonic forms and some nektonic and planktonic forms add to reef accumulations. Popular opinion assigns reef formation to corals, but it is probable that there are few existing reefs to which the corals made the major contribution. Corals held fourth place in the reef materials derived from the drilling on Funafuti; *Lithothamnium*, *Halimeda*, and foraminifera holding first, second, and third places in the order named. Corals made no contributions to the pre-Cambrian and Cambrian reefs. They became contributors in the Ordovician and have been prominent since. They may have been the chief contributors in the Silurian, but Cumings (1932) expressed the opinion that many of the Silurian reefs of Indiana had algae as the chief contributors. In general, organisms that participated in the formation of reefs may be placed in two classes, namely, those that constructed the framework and those that made additions by assisting in filling the spaces within the framework. To the former belong the colonial Hexacoralla, Tetracoralla, *Favosites*, *Halysites*, *Syringopora*, Heleolitidae, colonial Hydrozoa, calcareous algae, some bryozoans, and some sponges. To the second group belong the non-colonial coelenterates, foraminifera, crinoids, cystoids, brachiopods, mollusks, *et cetera*. Crinoidal remains are extremely abundant around some Silurian reefs and are held responsible for some bioherms in the Mississippian rocks.

REEFS IN GEOLOGIC COLUMN

Reefs seem to have been constructed somewhere in every geologic period since and including the Huronian and it is possible that some were built in pre-

Huronian time. They have been built in both fresh and salt water, but most are believed to be marine in origin. So far as known, temperatures for reef-building by corals must never fall below freezing, and for recent reefs the lower temperature limit for existence of the reef-building forms is about 68°F. Whether this was the lower temperature limit for the ancient reef-building forms is not known, but it is probable that the temperature permitting survival was not greatly different. The situation is different for algae. They thrive abundantly in the cold waters of the North Atlantic and they probably lived in cold waters of all geologic periods. Moreover, algae thrive in fresh as well as in saline water and probably have so thrived since their beginning.

Nearly half a century has passed since Grabau (1903) described the organic reefs of the Paleozoic and more than a third of a century (1913) since he described the reefs in the geologic column. Much more is now known of the occurrence of organic reefs in the rocks of the several eras.

PRE-CAMBRIAN REEFS

Reefs built during the pre-Cambrian eras were made by algae. Walcott has described algal deposits in the Huronian of the Grand Canyon region,[3] Montana, and British Columbia, and he states the presence of reefs formed by calcareous algae, and that such extend through many feet of limestone. He did not describe any pre-Cambrian reefs and it is probable that most if not all the algal deposits that he studied were distributed through beds of limestone and were biostromes and not bioherms. Fenton and Fenton found two reefs made by algae in the Belt series of Montana in the Granite Park member of the Siyeh formation. The constructing organism was *Collenia frequens*, a species described by Walcott. The reefs were found on the Going-to-the-Sun Highway in Glacier National Park. Although only two reefs were found, it is significant that both are exposed in artificial excavations so that it is very probable that many are present. One reef has an observed height of 25 feet with the base not exposed and a width of 33 feet. Dips of marginal limestones range to 20° and in the immediate forereef to 50°. Subjacent strata dip under the reef from 5° to 30°. The second reef has a height of 15 feet and is 21 feet wide. The base is exposed but the top has been removed by erosion. In cross section this reef has the shape of an inverted cone and enlarges upward so that the greatest width is at the top of the reef that remains. Both reefs seem to have been built on soft sediments into which they settled as they grew.

The lower Huronian of Upper Michigan contains an abundance of algal accumulations in the Kona, Randville, and Bad River dolomitic limestones. Many beds in each limestone are filled with the algal domes and some are composed almost entirely of the algal growths. Reefs do not seem to have been gen-

[3] Correlation of formations over any considerable distance in the pre-Cambrian is very uncertain and correlation of stratigraphic breaks is equally so. As this is the case, the writer is making the assumption that all occurrences of algal reefs in the pre-Cambrian are in the Huronian.

erally built and most accumulations are of the nature of the biostromes of Cumings. However, the algal structures in exposures of the Kona dolomitic limestones south of Marquette, Michigan, form a real reef. With the assistance of W. O. Hotchkiss this reef was measured in the spring of 1917 and was found to have a thickness of 22 feet with neither base nor summit exposed and a width of 55 feet with neither side shown. It may be much larger than the figures suggest. No reef accumulations were found in any of the exposures of the Randville and Bad River formations. This is probably due to the few exposures of these limestones. Algal accumulations are also present in the pre-Cambrian on the east side of Hudson Bay, but it does not seem to have been stated that reefs are present.

Schnock states that many algal reefs are in the pre-Cambrian formations of the Belgian Congo and that these were formed by algae of true *Collenia* and *Conophyton* types. Some colonies are reported to be 2 meters (6.5 feet) high and 4 meters (13 feet) wide.

CAMBRIAN REEFS

Algae and the group of organisms known as Pleospongia formed reefs in the Cambrian period. The Pleospongia built large accumulations in Labrador, Australia, and elsewhere in the Lower Cambrian and some of these are in the form of reefs. Reefs composed of the exoskeletal matter of Pleospongia have splendid development in the Lower Cambrian in Labrador on the Strait of Belle Isle. According to Schuchert one reef has a thickness of 50 feet, but the writer found a place where the thickness was between 75 and 100 feet.

Reefs of algae have been described from the Cambrian, particularly the Upper Cambrian, but where these have been seen by the writer the occurrences are biostromes rather than bioherms. *Cryptozoon proliferum* is present in abundance in the Hoyt limestone of New York and some of the occurrences may be in the form of reefs, but most seem to be biostromes. Blackwelder described "a fully exposed reef of calcareous algae?" in the Teton Mountains of Wyoming, but his description suggests a biostrome and not a bioherm. Derville has described reefs of algae in the Cambrian in the Carteret region (Manche) of France.

Since algae have rather common occurrence in Cambrian limestones, it seems likely that bioherms due to this group of organisms should be present.

ORDOVICIAN REEFS

Small reef-like bodies composed of corals and algae are present in the Chazy of Vermont and have been described by Raymond as "The oldest coral reef." The reef is stated to be nearly ½ mile long, at least 7 feet thick, and to have been constructed by a favositid coral known as *Lamottia* and the hydrocoralline *Stromatocerium* together with bryozoans and pelmatozoan echinoderms.

Reef-like bodies built by algae are common in North America in the Lower Ordovician dolomitic limestones generally known as the Beekmantown. They are very abundant in the Prairie du Chien formation of the Upper Mississippi Valley, the Romaine formation of the Mingan Islands on the north side of the Gulf of St. Lawrence where they form a small reef on Eskimo Island, in the Beekmantown limestones of the Lake Champlain region, and equivalent strata elsewhere. Most occurrences are of the nature of biostromes which extend through many feet, but some consist of domal stratification with the algae plastered one over another so that stratification is by no means regular. These are essentially reefs.

Small reefs built largely by corals are common in the Ellis Bay formation (High Cincinnatian) on Anticosti Island in the Gulf of St. Lawrence. The reefs are not large and were constructed by the hydroid corals *Beatricea* and *Clathrodictyon*, the favositids *Paleofavosites* and *Halysites*, and the heleolitids *Plasmopora* and *Lyellia*.

Hadding suggests (1926, p. 33) that some parts of the *Leptaena* limestone of Sweden are in the form of reefs and he notes particularly an occurrence in the quarry at Kalholn in the Dalecarlia region. The writer has seen the exposures in this quarry and questions this interpretation of any part of the formation exposed there in 1914.

SILURIAN REEFS

Reefs are abundant in the Silurian formations over most if not all of the world and this period probably represents the climax of the reef-building corals of the Paleozoic era. Some of the reefs are large and distribution is extensive. With interruptions they fringe the Michigan basin where Cumings and Shrock state (1928, p. 599) they have a linear distribution of 1,000 miles. Many Silurian reefs are crowded with corals of the genera *Favosites*, *Halysites*, *Heleolites*, *Plasmopora*, *Clathrodictyon*, and *Syringopora*. Many reefs are almost barren of fossils. Either the fossils were destroyed by diagenesis or the reefs were built by algae. A large dolomitized reef is exposed in the old Schoonmaker quarry at Wauwatosa, Wisconsin (Chamberlin, 1877). This reef is margined by more or less steeply dipping well stratified dolomitic limestones which originally may have been calcitic sands and muds. These are draped around the exposed part of the reef like a shawl and ultimately pass over it in the form of a dome or anticline. An apparent closure of about 60 feet (Shrock, 1939, gives 63 feet) exists in the exposed reef as the reef is that high. As neither the base of the reef nor the distance the beds dip away from it is exposed, the closure is certainly much greater and the reef may be fully 100 feet thick. South of Fond du Lac in the Hamilton quarries the strata overlying reef structures are exposed. These consist of thin slabby limestones in highly billowy attitudes with small domes, basins, anticlines, and synclines. Some domes and anticlines have closures of 10–20 feet. These overlying strata belong to the Byron formation and the reefs are based in the top of the Mayville formation and extend into the Byron. This is one of the finest exposures showing the relation of reefs to the formation of large sedimentary structures that the writer has ever seen. An excellent exposure of a reef is in the quarry of the Union Lime Company at the village of Grimm. This reef is 16 feet thick and fully 100 feet wide. The quarry in the town of Cedarburg exposes a reef that is 30 feet high, and one in the quarry at Grafton is 25 feet high. There are many others and they may be found in quarries in eastern Wisconsin almost to the end of the Green Bay Peninsula, but seem to be most common south of the town of Green Bay.

The organic reefs of Indiana have been excellently described by Cumings and Shrock. They state that more than 70 individual reefs are known and that they extend over an area 60 miles north and south and 90 miles east and west. The heights range to 100 feet and widths from a few hundred feet to more than 1 mile. They also note the occurrence of reefs in Ohio and Iowa, and also those in Wisconsin. A reef in a cut on the Big Four Railroad at Wabash, Indiana (1928, p. 601), has a core that is 250 feet in diameter which consists of unstratified porous dolomitic limestone with little pockets and specks of calcite (Fig. 2).

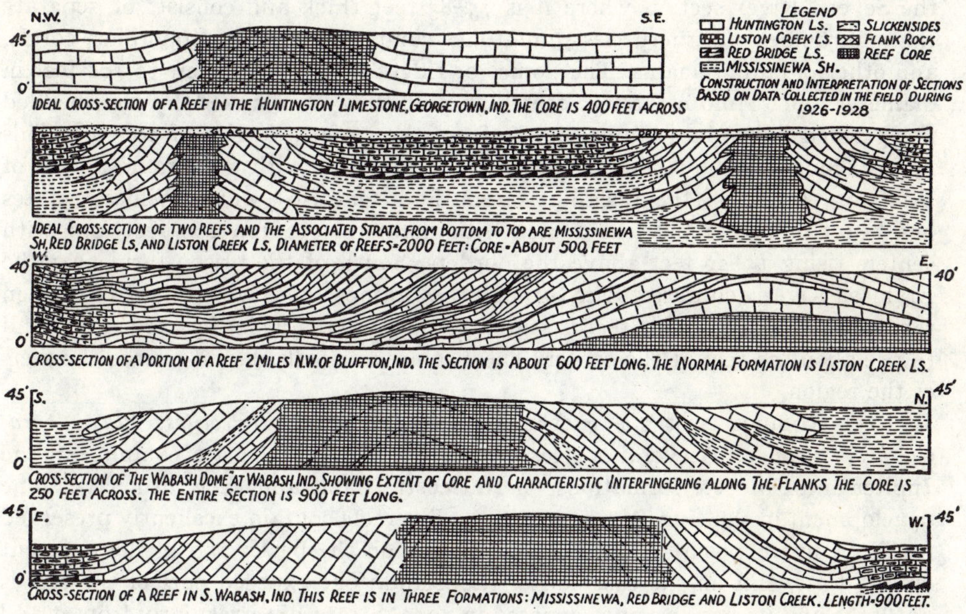

FIG. 2.—Diagrams of reefs in Indiana, after Cumings and Shrock, 1928.

These were identified as diagenetically modified remains of stromatoporoids. Most of the core is stated to have been originally a breccia composed of stromatoporoid and other organic remains in which few fossils could be identified. This breccia may have been produced by destructive organisms. Dipping away from the core are stratified dolomitic limestones composed of materials washed from the core and remains of organisms that lived on the sloping bottom. These rocks represent the forereef. Dips are stated to range to 65°. The stratification of the rocks of the forereef is fair, beds are thick and thicken and thin abruptly and interwedge outward into the inter-reef deposits which originally consisted of calcareous muds.

Many of the reefs of Indiana which have what has generally been termed stromatoporoid structure are considered by Cumings to have been built by algae (Cumings, 1932, p. 340). It is considered probable that reefs in the Silurian of Wisconsin which seem to be non-fossiliferous were also built by algae.

It is thought that reefs are present in the Silurian strata that underlie the Michigan basin and that drilling may be expected to encounter some of them.

Savage and Van Tuyl have described reefs in the Silurian of the Hudson Bay region. The Attawapiskat coral reef in the Ekwan River section has a thickness of 30 feet and is composed of domes and ridges of stromatoporoids and corals flanked by stratified fine-grained limestones. The reef is underlain by 24 feet of limestone largely composed of domes of corals separated by fine-grained limestones inclined at various angles. The Attawapiskat coral reef is also exposed in the Severn River section where it is 75–85 feet thick and consists of separate ridges or domes of structureless limestones composed of stromatoporoids, corals, and other organic remains. The domes or ridges range from a few to 100 feet in diameter and to more than 30 feet high and they are flanked by thin-bedded fine-grained limestones dipping away from the reefs at angles of 20°–30°. This reef is thought to be Niagaran in age and correlative with the Eramosa beds of Williams—a part of the Lockport of southern Ontario—which Williams states has beautifully symmetrical domes with diameters of 100–200 feet and with centers rising 15–20 feet above the borders. South of the prison farm near the Eramosa River a coral reef rises through the Eramosa beds which there have been removed by erosion. This reef is stated to be 35 yards wide, 85 yards long, and to rise about 20 feet at the center. The Eramosa beds flank and overlie the reefs of the region.

Reefs composed of the exoskeletons of *Halysites*, *Heleolites*, *Plasmopora*, *Clathrodictyon*, and several species of colonial Tetracoralla are common in the Jupiter and Chicotte formations on Anticosti Island and they also have some development in the Gun River formation. The reefs contain excellently preserved fossils. A fine exposure of a reef may be seen on the beach and in the adjacent cliff at East Cliff on the east end of the island. The exposed part of the reef is not large and its base was not exposed in 1919. Strata dip irregularly from it and ultimately arch over it. The reef extends out into the water and in the days before the island had lighthouses it was a menace to ships. The steamship Circe was wrecked on this reef in the early nineties and two of the boilers of this ship lay on the reef as late as 1919. The downward extent and linear distribution of this reef are not known. The Chicotte formation has many reef-like bodies around some of which there seem to have been gardens of crinoids whose stems are present in extremely great abundance; crinoid calices are rare. These were important contributors to the reef accumulations. Some beds of the Chicotte formation are very largely composed of crinoid stems and segments and in places the entire formation seems to be composed of a structureless mass of *Clathrodictyon*, *Favosites*, *Heleolites*, *Plasmopora*, *Halysites* and colonial forms of Tetracoralla together with other fossils.

The little altered reefs of the Island of Gotland in the Baltic Sea have been studied for more than a century and the steep dips in connection with the reefs have occasioned much interest and some misinterpretation of the structure of the island. Steep dips may be seen in many places on the island and are excellently shown on the small island of Storo Karlso on the west side of Gotland. This small island has strata dipping west on the west side and east on the east side, both from a central core. The coral *Omphyma* is very abundant in connection with this reef. The most splendid exposure of the core of a reef and the steeply and irregularly dipping flanking strata is in the Hobergen Klint on the south end of the island. The core of this reef is 150 yards in diameter (Fig. 1) and it is about 75 feet high. One stromatoporoid in this reef that was measured by the writer in 1914 was found to have a diameter of $10\frac{1}{2}$ feet and a thickness of 8 feet. Many high places in the interior of Gotland, known as *Klintar*, have reef cores which are responsible for the elevations. These have been favorite sites for the high-steepled churches that dot the Gotland landscape. Corals are extremely well preserved in the reefs of Gotland and clayey patches between the coralline growths generally contain excellently preserved fossils of other groups. The reefs have been studied over the past 50 years by Munthe, Hedstrom, Hadding, Maillieux, Wiman, and others. Hadding states that the reefs are at different horizons and the formation of a reef was completed at one place before another was started at another place. This statement requires some modification. The rocks of the reefs are described by Hadding as structureless with some parts porous, others dense, and with the marginal rocks inclining in all directions from the reefs. Some reef rocks are stated to be fine-grained, others coarse-grained.

There are also reefs in the Silurian of Esthonia on the east side of the Baltic Sea. A small reef was exposed in 1914 near Weissenstein on the estate owned at that time by the Munterhoff family and a large reef was exposed at that time in the sea cliff south of Hapsal near the village of Pasko. These places probably now have different names. So far as the reefs of Esthonia were seen by the writer, they have been rather completely recrystallized and many are coarse-grained crystalline limestones with only remnants of corals preserved. Many fossil corals are represented by cavities whose origins are indicated by fragments of coralline structure on the margins.

The Upper Wenlock of England contains bodies of unstratified limestone. These are known as ballstones. They have lenticular to ovoidal shapes in vertical section and range to 90 feet thick. Marginal strata dovetail into the ballstones, end abruptly against them, and ultimately arch over them. The structures contain favositid, stromatoporoid, and cyathophyllid corals of which 96 per cent are in positions of growth, whereas in marginal strata only 16 per cent are in positions of growth. These seem to be reefs that grew under conditions not favorable for lateral expansion.

DEVONIAN REEFS

Coral reefs have been reported from the Devonian of New York, Michigan, and Indiana. Most of these seem to be of the nature of biostromes rather than bioherms (Cumings, 1932, p. 545). The frequently mentioned coral reefs in the Devonian in the falls of the Ohio River in Louisville, Kentucky, seem to be more like beds filled with corals than a reef, and according to Cumings they do not swell into true reef form. North, in his book on limestones, cautions that mere abundance of corals in a formation does not justfy terming the deposit a reef. However, since corals are abundant in the Devonian it would be indeed strange if they did not form reefs over shallow bottoms where conditions were favorable. Grabau figured and described a reef in the Onondaga formation at Williamsville, New York, which he stated is several hundred feet in diameter and into which marginal strata interfinger and from which they dip at angles up to 10° (1913, pp. 425–26). The corals in the reefs are stated to be branching forms of *Syringopora*, *Eridophyllum*, *Synaptophyllum*, and *Diplophyllum*. *Favosites* is said to be common. Grabau also figures a reef in the Devonian Alpena limestone of Michigan (1913, p. 427) composed of stromatoporoids, *Favosites*, *Acervularia*, and other colonial forms. Many colonies are 3–4 feet in diameter and a few stromatoporoids are 10 feet in diameter. On the Lake Michigan side there are also large reefs in Devonian strata and it is very probable that reefs are present throughout the Devonian strata in Lower Michigan. Drilling seems to have found some of them. Schuchert and Dunbar (1941, p. 207) figure a small coral reef in the Alpena limestone and they state that reefs to more than 50 feet in diameter are present in that limestone. Smith states that the Alpena limestone is characterized by extensive coral reefs and Cooper gives the information that the Alpena limestone as restricted by Warthin and Cooper is 79 feet thick and contains large coral bioherms in the upper 40 feet. It is probable that bioherms are as common in the subsurface as in the exposures.

Bell figured and described extensive reefs to which corals were contributors on the Attawapiskat River, a stream flowing into James Bay, in strata which he identified as Devonian.[4] The reefs were described as spongy masses of limestone without stratification and some were at least 40 feet thick. Brachiopods were reported to be the chief contributors. The reefs were stated to be margined by thin-bedded limestones which dip away from the reefs at various angles. In the recently discovered Leduc oil field in Alberta, oil is produced from a dolomitic limestone identified as Upper Devonian from the two depths of 5,000 and between 5,150 and 5,400 feet. The consensus favors the view that the accumulation is in an Upper Devonian fossil reef. This is the view advanced by Link and Layer and collaborators. The Devonian strata are divided by Layer *et al.* into three zones: D1, D2, and D3. The D3 zone seems to be the reef. This has been completely dolomitized and is without bedding or has inclined bedding as should be characteristic of marginal strata. Porosity seems connected with dolomitization and probably leaching. Link suggested that the Norman Wells oil reservoir is a Devonian reef. Savage and Van Tuyl described Devonian strata in the Hudson Bay region and gave a section exposed on the Abitibi River. They state that zone

[4] The writer is not certain that the strata in which the reefs occur are Devonian in age.

8 of their section contains a coral reef in the lower 3 feet. This is not a large reef.

Splendid reefs have been described as present in the Middle Devonian in the Eifel region of Germany and the Ardennes region of France and Belgium. The formation in the Eifel region is known as the Eifel limestone and its equivalent in the Ardennes region is known as the Givitien formation. This formation has a maximum thickness of 1,200 feet and contains an abundance of corals and coral reefs. As in the Silurian of Indiana and Gotland, the reefs are responsible for some of the scenery. The chief contributors to the building of the reefs were favositid, cyathophyllid, and stromatoporoid corals (Dupont). The reef materials are greatly altered so that fossils are not readily apparent. Reefs of organic origin are also present in the Karnic Alps where they are composed in part of favositid, stromatoporoid, and other corals together with remains of other organisms.

MISSISSIPPIAN REEFS

It does not seem to be known that algal or coral reefs existed in North America during the Mississippian period. The extensive deposition of calcareous sediments over the Mississippi Valley and in the Rocky Mountain region seems to have taken place in an environment not favorable to the growth of corals, but *Lithostrotion* is a recognized colonial form and reefs should be expected in the rocks. Why algae are not common is not known and it is probable that they are present but have not been recognized. Algal reefs may be expected to be present.

Stockdale has described bioherms in the Mississippian Borden group of Indiana. These consist of stratified limestone masses of reef-like form surrounded by shales and sandstone. One illustrated by Stockdale is the Floyd Knob-Edwardsville bioherm in southeastern Monroe County. This has a basal diameter not exceeding 2 miles, is based in the Floyd Knob formation, rises into the Edwardsville formation, and has a total thickness exceeding 60 feet. The limestones in the bioherm are stratified, surrounding materials do not dip away from it, and the composing limestones are dominantly crinoidal, but beds of shale are interstratified.

Laudon and Bowsher state that the Alamogordo member of the Lake City formation of the Mississippian system of New Mexico as exposed in the Sacramento Mountains contains bioherms of structureless limestones composed of crinoidal remains. The central part of a bioherm has a distinct absence of bedding, but bedding becomes recognizable toward the margins. The bioherms range from small bodies to a mile in diameter and 200 feet in thickness. Beds dip steeply from the bioherms with a range to 40°. Similar structures were found in the same member of the Lake City formation in southwestern New Mexico. Laudon also notes the occurrence of small bioherms in the St. Joe formation of northeastern Oklahoma.

Dupont states that the Calcaire de Waulsort of Belgium is a massive structureless limestone largely composed of hydrocorallines and that corals play a subordinate part. The reefs are described as small massive heaps scattered through the limestones. One reef is said to have a continuity of 60 kilometers (a little more than 37 miles). Most reefs are small domal structures surrounded by well bedded limestones.

According to Parkinson, reef limestones in the Viséan (upper Mississippian) of Lancashire and Yorkshire, England, accumulated at high angles to the horizontal and at depths ranging from sea-level down to 400 or more feet. The structures interpreted as reefs are circular to ovoidal in plan and rise as masses 300–400 feet in the associated rocks. Thin-bedded limestones and shales are stated to have been contemporaneously deposited between the reefs (Tiddeman). The supposed reefs are stated to be without bedding (Dakyns) and are interpreted as formed by accumulations of shells. Marr does not agree with this interpretation and believes the supposed reefs as due to tectonic movements.

PENNSYLVANIAN REEFS

The rocks of the Pennsylvanian system contain many biostromes as shown by the many zones filled with brachiopods, *Myalina*, and foraminifera. Algae are abundant in some formations, small bioherms are recognized, and Milburn Stryker and the writer mapped a small one in southeastern Kansas in 1915. The rarity of bioherms in the Pennsylvanian strata of interior North America may be explained as due to the large quantities of muddy water brought into the interior seas.

Schuchert and Dunbar (1941, p. 259) figure after Holtedahl a coral reef in early Pennsylvanian strata (Moscovian) in western Spitzbergen. The high latitude of this reef strongly suggests that reefs should have been common in Pennsylvanian strata in regions nearer the equator. It may be found that reefs are common in thick Pennsylvanian limestones.

PERMIAN REEFS

Some of the largest fossil organic reefs in the geologic column are in the Permian system where they are well known in the Trans-Pecos region of Texas and New Mexico and in southeastern Russia. According to P. B. King, the reefs of the Trans-Pecos region are massive structures built by algae, corals, crinoids with large columns, and thick-shelled brachiopods and brachiopods with high and deformed cardinal areas. King states that the relative abundance of the different kinds of organisms in the associations varies from place to place and no one group can be considered the chief builder. Some parts of the reefs contain few determinable organic remains, a fact correlatable with possible destruction by boring organisms, the obvious diagenesis that led to the dolomitization of the originally calcitic sediments, and the erosion to which the reefs must have been subjected. Reefs are present in both the lower Permian (Word formation) at the higher Permian (Capitan limestone). King suggests that in the lower Permian the reefs rose about 100 feet above the bottom, but in the higher Permian they had a relief above the bottom from 300 to more than 500 feet. The reefs are stated to resemble those of recent seas in having great linear extent and limited width and are not like the somewhat circular bodies in the Silurian of Indiana, Wisconsin, Anticosti, Gotland, and Esthonia. Massive dolomitic limestones dip away from the reefs with inclinations of 20° or more. According to King the reefs were constructed at a greater distance from land than are recent reefs.

The organic reefs in the Permian of southeastern Russia are in the lower part (Sakmarian series) and some form high and isolated hills from which erosion has removed the original bordering strata. These hills are known as Shikhans. Dunbar figures the Shikhan Tra-tau which rises about 900 feet above the surroundings and he suggests that its base is much deeper than appears from the part exposed and that the thickness of the reef may be greater than 3,000 feet. The strata marginal to the reef are the Artinskian limestones, but the reef is believed to be based in lower strata with which the reef materials are probably equivalent. The exposed part of this reef covers about 1 square mile and algae seem to have been the chief builders.

A few miles north of the Shikhan Tra-Tau is the Shikhan Trak-Tau. This reef is elongate and seems to be part of a partly buried reef system. If Dunbar's suggested thickness of the Shikhan Tra-Tau is approximately correct, this is the thickest known reef in the geologic column.

Reefs are reported from the Zechstein of Germany (Magdefrau) which are largely composed of bryozoans with the next most abundant organism either a hydrocoralline or a sponge. There may be organic reefs in the Permian of Asia, Australia, and Africa, but the writer has no information.

TRIASSIC REEFS

Muller states that coral reefs occur in the Triassic of Nevada (Dunlap Canyon) and that the reef-building corals resemble those in the Triassic of northern California. No details of the reefs are given.

The Schlern dolomitic limestones of South Tyrol were originally interpreted as a great barrier reef system (Mojsisovics, 1879) with bordering strata dipping steeply from the margins. Ogilvie-Gordon questioned this origin of the steep dips and considered these features as due to faulting and erosion. Mojsisovics and his contemporaries thought the supposed reefs were built by corals, but Rothpletz and von Gümbel showed that algae were the chief builders of the rocks although other organisms contributed. Ogilvie-Gordon stated that the reefs of earlier students were masses of dolomitic limestone isolated by faulting and erosion and that the real reefs were small and were built by corals and echinoderms together with volcanic ash and clastic sediments. Algae made deposits in shallow and protected waters behind reef barriers. Progressive subsidence was essential for the accumulation of 3,000 feet of limestones of which considerable is of algal origin, and this indicates that a reef barrier must extend through nearly the same thickness, probably varying in extent from time to time and it is not unlikely that at times the coral and echinoderm reefs became overgrown by the algae.

JURASSIC REEFS

The Jurassic of Europe contains many bodies of reef character and in England a division is known as the Corallian by reason of the abundance of corals. Arkell has described broad but thin accumulations in the Jurassic which he terms reefs. The descriptions suggest that many, perhaps all, are biostromes. The lower part of the Jurassic of the Swabian Alps of southern Germany contains structureless reef bodies composed of sponges. These are margined by, and interfinger with, stratified limestones. The higher parts of these limestones contain small units composed of shells and calcareous algae.

Perhaps the most famous reefs in the Jurassic are those in the region of Solnhofen in Bavaria. The rocks of these reefs are structureless dolomitic limestones whose building organisms consist of sponges, corals, and bivalve mollusks, the last being very abundant in some places. The reefs are margined by thin-bedded fine-grained limestones which are famous because of their use in lithographic work. These are the deposits of the lagoons that lay between, and in some places finally over, the reefs. The rocks of the reefs seem structureless and interfinger with the marginal fine-grained limestones.

Reefs in the so-called White Jura are stated by Berckheimer to have the hydroid coral, *Ellipsactinia*, as the chief builder. Also, in the White Jura are massive structureless masses which are referred to the work of sponges (Roll). Bigot stated the occurrence of organic reefs in the Bathonian of Normandy, and Freire de Andrade found them in the Jurassic (Lusitanian) of Portugal.

CRETACEOUS REEFS

Few Cretaceous reefs seem to be known in North America. The tepee buttes of Colorado, composed of accumulations of shells in columnar form, seem to be the nearest things to reefs that are known. In the Mediterranean region of Europe structureless bodies of limestone composed of shells of rudistid pelecypods together with algae, foraminifera, and other organic materials form reef-like structures. These are surrounded by stratified Lower Cretaceous limestones. According to Hadding, coral reefs are present in the Cretaceous (Danian) of southern Sweden. These are not large. In view of the great extent of limestones in the Cretaceous, it is strange that reefs are not common. Perhaps the fine-grained character of the lime sediments was not favorable for the growth of reef-building organisms.

TERTIARY REEFS

The writer is not aware that coral reefs are exposed at the surface in North America. They are not common in Europe. Andrussow has described reefs built by bryozoans in the Pliocene at Akra in Crimea. These reefs are not surrounded by limestones. Heights are stated to range to 30 meters and sides are very steep. Dangeard states that coral reefs are present in the Oligocene on the Limagne border of the Central Plateau of France. Miocene reefs built by corals and *Vermetus* are reported to be present in Galicia.

Bradley has described reefs built by algae in the Green River formation of Colorado, Utah, and Wyoming. Johnson states the occurrence of algal reefs in Oligocene lakes in the South Park region of Colorado, and Price, Elias, and Frye have described reefs of the same origin in the Ogallala formation in the Llano Estacado of New Mexico and Texas. Elias has noted similar reef-like forms in the same formation of western Kansas.

PLEISTOCENE REEFS

It is probable that some of the reefs in present seas began their development in Pleistocene time. This is difficultly determinable. In the United States it is clear that the tufa deposits in Pleistocene Lake Lahontan are forms of reefs. These seem to have been built by algae. The deposits are of various shapes and dimensions and range from bedded deposits in the form of biostromes to structures which have typical reef-like or biohermal form.

SUMMARY

The foregoing pages show that organic reefs or bioherms are present in every system beginning with the Huronian. In some systems they extend to polar latitudes and the occurrences indicate a wider subsurface distribution. Reefs may be encountered in every extensive calcareous formation.

Reefs may have original porosity in three places: the reef itself, the flanking strata, and the overlying strata. Diagenetic processes, among which is dolomitization, and the work of boring organisms may increase or decrease porosity and permeability. Cementation may be so complete that porosity and permeability may be almost eliminated. Unaltered reefs may have large porosity and little permeability. Altered reefs may have a range in porosity and permeability from very little to very much. Both porosity and permeability may greatly vary from place to place and from time to time. The extent of porosity and permeability can not be forecast in advance of drilling.

Heights and extents of ancient reefs are of great variation. Most reefs in early and middle Paleozoic seem to have been somewhat circular in plan and of no great height, but the reefs might have been distributed over a great area. Beginning in the Permian, large reefs developed and some of these had great linear extent and thickness. Reefs have little known occurrence in the Mesozoic of North America.

BIBLIOGRAPHY

ANDRUSSOW, N., *Die fossilen Bryozoenriffe der Halbinseln Kertsch und Taman*, Lieferungen 1–3 (1910–1912). Kief.

———, "Vergleich der fossilen Bryozoenriffe der Halbinseln Kertsch und Taman mid anderen riffeartigen Bildungen," *Bull. de l'assoc. Russe pour les recherches scientifiques à Prague*, Vol. 4 (1936).

ARKELL, W. J., *The Jurassic System of Great Britain*, p. 615. London (1933).

———, "On the Nature, Origin and Climatic Significance of the Coral Reefs in the Vicinity of Oxford," *Quar. Jour. Geol. Soc. London*, Vol. 91 (1935), pp. 77–110.

BELL, R., "Reef Structures on the Attawapiskat River," *Canada Geol. Survey Ann. Rept.*, Vol. 2 (1886), pp. 27G–28G.

Berckheimer, F., "Eine vorläufige Mitteilung über den Aufbau der weissen Jura in Schwaben," *Jahreshefte der Vereins für Vaterl., Naturkunde in Wurtemburg*, Jahrg. 69 (1913), pp. lxvi–lxxii.
Blackwelder, E., "A Fully Exposed Reef of Calcareous Algae (?) in the Middle Cambrian of the Teton Mountains," *Amer. Jour. Sci.*, Vol. 39 (1915), pp. 46–65.
Bradley, W. H., "Algal Reefs and Oölites of the Green River Formation," *U. S. Geol. Survey Prof. Paper 154G* (1928).
Chamberlin, T. C., "Geology of Wisconsin," *Wisconsin Geol. Survey*, Vol. 2 (1877), pp. 360–71.
Cooper, G. A., "Correlation of the Devonian Sedimentary Formations of North America," *Geol. Soc. America*, Vol. 53 (1942), p. 1742.
Cumings, E. R., "Reefs or Bioherms?," *Bull. Geol. Soc. America*, Vol. 43 (1932), pp. 331–52.
———, and Shrock, R. R., "The Silurian Coral Reefs of Indiana and Their Essential Strata," *Proc. Indiana Acad. Sci. 1926* (1927), pp. 71–85.
———, "Niagaran Coral Reefs of Indiana and Adjacent States and Their Stratigraphic Relations," *Bull. Geol. Soc. America*, Vol. 39 (1928), pp. 579–620.
Dakyns, J. R., "The Limestone-Knolls below Skipton and Grassington in Craven," *Quar. Jour. Geol. Soc. London*, Vol. 55 (1899), 359–61.
Dangeard, L., "Quelques observations sur le Tertiare de la Limagne Bourbonnais; importance des phénomènes karstiques dans à la présence des calcaires concrétionnes," *Bull. Geol. Soc. France*, Ser. 5, T. 3 (1934), pp. 381–84.
Derville, E., "Les récifs en coupole des calcaires Cambriens de la région Carteret (Manche)," *Compt. Rend. Acad. Sci. Paris*, T. 198, No. 6 (1934), pp. 588–90.
Dunbar, C. O., "The Type Permian: Its Classification and Correlation," *Bull. Geol. Soc. America*, Vol. 24 (1940), pp. 254–58.
Dupont, E., "Sur l'origine des calcaire Dévoniens de la Belgique," *Bull. l'Acad. Roy. Belgique*, Ser. 3, T. 2 (1881), pp. 264–80.
———, "Sur les origines du calcaire Carbonifère de la Belgique," *ibid.*, Ser. 3, T. 5 (1883), pp. 211–29.
Elias, M. K., "The Geology of Wallace County, Kansas," *Kansas Geol. Survey Bull. 18* (1931), pp. 136–40.
Endo, R., "Method of Boring and Testing the Formations on Kita-Daito-Zima," *Geol. and Paleon. Inst. Tohoku Imp. Univ.*, No. 14. Cited by Ladd and Tracy.
Fenton, C. L., "Niagaran Stromatoporoid Reefs in the Chicago Region," *Amer. Midland Nat.*, Vol. 12 (1931), pp. 203–12.
———, and Fenton, M. A., "Algal Reefs or Bioherms in the Belt Series of Montana," *Bull. Geol. Soc. America*, Vol. 44 (1933), pp. 1135–42.
Freire de Andrade, C., "Um recife corálico no Jurássico de Vila Franca de Xira," *Bull. Mus. e Lab. Miner. e Geol.*, Ser. 1, No. 3 (1934), pp. 189–97.
Gilbert, G. K., and Gulliver, F. P., "Tepee Buttes," *Bull. Geol. Soc. America*, Vol. 6 (1895), pp. 333–42.
Grabau, A. W., "Paleozoic Coral Reefs," *Bull. Geol. Soc. America*, Vol. 14 (1903), pp. 337–52.
———, *Principles of Stratigraphy* (1913), pp. 384–445.
Gray, J. L., and Fournier, F. L., "Developments in Canada in 1947," *Bull. Amer. Assoc. Petrol. Geol.*, Vol. 32, No. 6 (1938), pp. 1081–83.
Gümbel, C. W. von, "Geognostische Mittheilungen aus den Alpen, Das Mendel- und Schlerngebirge," *Sitz. der Akad. f. Wiss. in Munchen, Math. Phys. Klasse* (1873).
Hadding, A., "The Pre-Quaternary Sedimentary Rocks of Sweden," *Med. Lunds Geol.- Min. Inst.*, No. 32 (1926), p. 33.
Hedstrom, H., "The Silurian Stratigraphy in the Neighborhood of Visby," *Guide to Excursions, Internat. Geol. Congress*, 11th session, Stockholm (1910).
Hoffmeister, J. E., and Ladd, H. S., *Jour. Geol.*, Vol. 52 (1944), pp. 388–402.
Johnson, J. H., "Algal and Algal Limestones from the Oligocene of South Park, Colorado," *Bull. Geol. Soc. America*, Vol. 51 (1937), pp. 1227–36.
Karlov, N. N., "On the Age and Conditions of the *Membranipora* Reefs in the Kersch Peninsula," *Acad. Sci. U.S.S.R.*, Geol. Ser. (1937), pp. 1003–36.
King, P. B., "Permian Stratigraphy of Trans-Pecos, Texas," *Bull. Geol. Soc. America*, Vol. 45 (1934), pp. 697–798.
Kuenen, P. H., *Verh. Koninklijke Akad. Wetenschappen*, Pt. 43, No. 2 (1947), pp. 1–35. Cited by Ladd and Tracy.
Ladd, H. S., and Tracy, J. L., "Drilling on Bikini Atoll, Marshall Islands," *Science*, Vol. 107 (1948), pp. 51–55.
Laudon, L. R., "Stratigraphy of the Osage Subseries of Northeastern Oklahoma," *Bull. Amer. Assoc. Petrol. Geol.*, Vol. 21 (1939), pp. 1158–67.

———, AND BOWSHER, A. L., "Mississippian Formations of Sacramento Mountains, New Mexico," *Bull. Amer. Assoc. Petrol. Geol.*, Vol. 25 (1941), pp. 2107–60.
———, "Mississippian Formations of Southwestern New Mexico," *Bull. Geol. Soc. America*, Vol. 60 (1949), pp. 1–88.
LAYER, D. B., AND MEMBERS OF STAFF, IMPERIAL OIL LIMITED, "Leduc Oil Field, Alberta, a Devonian Coral-Reef Discovery," *Bull. Amer. Assoc. Petrol. Geol.*, Vol. 33 (1949), pp. 572–602.
LINK, T. A., "Leduc Oil Field, Canada," *Bull. Geol. Soc. America*, Vol. 60 (1949), pp. 381–402.
MAGDEFRAU, K., "Zur Entstehung der mitteldeutschen Zechstein-Riffe," *Centralb. Miner.*, Abt. B. No. 11 (1933), pp. 621–24.
———, "Aufbau der thüringeschen Zechstein-Riffe," *Natur und Volk*, B. 67, G. 1 (1936), pp. 48–58.
MAILLIEUX, E., "Les récifs silurien de l'ile de Gotland," *Bull. Soc. Belge Geol.*, T. 43 (1933), pp. 206–08.
MARR, J. E., *Quar. Jour. Geol. Soc. London*, Vol. 55 (1899), pp. 327–58.
MOJSISOVICS, E., *Die Dolomit-Riffe von Südtirol und Venetien*. Vienna (1879).
MULLER, S. W., "Triassic Coral Reefs in Nevada," *Proc. Geol. Soc. America*, 1935 (1936), p. 410.
———, "Triassic Coral Reefs of Nevada," *Amer. Jour. Sci.*, Vol. 31 (1926), pp. 202–08.
MUNTHE, H., "The Sequence of Strata in Southern Gotland," *Guide to Excursions, Internat. Geol. Congress*, 11th session, Stockholm (1910).
NORTH, F. J., *Limestones, Their Origins, Distributions and Uses*. London (1930).
OGILVIE-GORDON, M. M., "Das Grödener-Fassa und Enneberggebiet in die Südtiroler Dolomiten," *Geologische Bundanstalt* (1927).
PARKINSON, D., "The Geology and Topography of the Limestone Knolls in Boland (Bowland), Lancashire and Yorkshire, England," *Proc. Geol. Assoc.*, Vol. 46 (1935), pp. 97–120.
———, "The Origin and Structure of the Lower Viséan Reef-Knolls in the Clitheroe District, Lancashire," *Quar. Jour. Geol. Soc. London*, Vol. 99 (1944), pp. 155–169.
PRICE, W. A., ELIAS, M. K., AND FRYE, J. C., "Algal Reefs in Cap Rock of Ogallala Formation on Llano Estacado Plateau, New Mexico and Texas," *Bull. Amer. Assoc. Petrol. Geol.*, Vol. 30 (1946), pp. 1742–46.
RAYMOND, P. E., "The Oldest Coral Reef," *14th Bienn. Rept. Vermont State Geologist for 1923–1924* (1924), pp. 72–78.
RICHARDS, H. C., AND HILL, D., "Great Barrier Reef Bores," *Rept. Barrier Reef Committee*, Vol. 5 (1926, 1937). Brisbane, Australia.
ROLL, A., "Form, Bau, und Entstehung der Schwamstotzen in süddeutschen Malms," *Pal. Zeitsch.*, B. 16 (1934), pp. 197–246.
ROTHPLETZ, A., *Ein geologischen Querschnitt durch die Ost-Alpen nebst Anhang, über de sogenannten Glarner Doppel-Falte*. Stuttgart (1894).
RUSSELL, I. C., "Geological History of Lake Lahontan," *U. S. Geol. Survey Mon. 2* (1885), pp. 189–222.
SAVAGE, T. E., AND VAN TUYL, F. M., "Geology and Stratigraphy of the Area of Paleozoic Rocks in the Vicinity of Hudson and James Bays," *Bull. Geol. Soc. America*, Vol. 30 (1919), pp. 339–78.
SCHNOCK, P., "Sur les Algues carcaires découtert dans le sud-est de la Cuvette centrale du Congo Belge," *Compt. Rendu Acad. Sci. Paris*, T. 224 (1947), pp. 1367–71.
SHROCK, R. R., "The Klintar of the Upper Wabash Valley in Northern Indiana," *Jour. Geol.*, Vol. 37 (1929), pp. 17–29.
———, "Wisconsin Bioherms," *Bull. Geol. Soc. America*, Vol. 50 (1939), pp. 529–62.
SCHUCHERT, CHARLES, AND DUNBAR, CARL O., *Historical Geology*. John Wiley, New York (1941).
SIEBER, R., "Weiters Ergebnisse paläobiologischer Untersuchungen an der Fauna der rhätischen Riffkakle der Rötelwand und anteren rhätischen der Nordalpen," *Akad. Wiss., Vienna, Math.-Natur. Kl.*, Vol. 71 (1934), pp. 268–61.
SMITH, R. A., "Limestones of Michigan," *Michigan Geol. Survey Pub. 21*, Ser. 17 (1916), pp. 103–11.
STOCKDALE, P. B., "Bioherms in the Borden Group of Indiana," *Bull. Geol. Soc. America*, Vol. 42 (1931), pp. 707–18.
TIDDEMAN, R. H., "On the Concurrent Faulting and Deposit in Carboniferous Times in Craven, Yorkshire, with a Note on Carboniferous Reefs," *Rept. Brit. Assoc.* (1889), pp. 600–03.
TWENHOFEL, W. H., "The Silurian and High Ordovician Strata of Esthonia, Russia, and Their Faunas," *Bull. Mus. Comp. Zool.*, Vol. 56 (1916), pp. 707–18.
———, "An Interpretation of the Silurian Section of Gotland," *ibid.* (1916), pp. 341–54.
———, "Pre-Cambrian and Carboniferous Algal Deposits," *Amer. Jour. Sci.*, Vol. 48 (1919), pp. 339–52.
———, "The Geology of Anticosti," *Canada Geol. Survey Mem. 154* (1927).
VAUGHAN, T. W., "Physical Conditions under Which the Paleozoic Reefs Were Formed," *Bull. Geol. Soc. America*, Vol. 22 (1911), pp. 238–52.

WALCOTT, C. D., "Pre-Cambrian Algal Flora," *Smithsonian Misc. Coll., Cambrian Geology and Paleontology*, Vol. 64, No. 2 (1914).

WALTHER, J., "Die Adamsbrücke und die Korallenriffe der Palkstrasse, Sedimenstudien in tropischen Litoralgebiet," *Petermann's Mit. aus J. Perthes Geog. Anstalt*, Erganzungband, 22, H. 102 (1891), pp. 1–40.

———, "Die Fauna der solnhofener Plattenkalke bionomisch betrachtet," *Festschrift zum siebzigsten Geburtstage von Ernst Haeckel*. Jena (1904).

WARTHIN, A. S., JR., AND COOPER, G. A., "New Formation Names in the Michigan Devonian," *Jour. Washington Acad. Sci.*, Vol. 25 (1935), pp. 524–56.

WILLIAMS, M. Y., "An Eurypterid Horizon in the Niagara Formation of Ontario," *Canada Geol. Survey Mus. Bull. 20* (1915).

Reprinted from
BULLETIN OF THE AMERICAN ASSOCIATION OF PETROLEUM GEOLOGISTS
VOL. 34, NO. 2 (FEBRUARY, 1950), PP. 215-238, 14 FIGS., 1 TABLE

CRETACEOUS AND TERTIARY REEF FORMATIONS AND ASSOCIATED SEDIMENTS IN MIDDLE EAST[1]

F. R. S. HENSON[2]
London, England

ABSTRACT

This paper makes no contribution to general theories of reef-formation or to detailed studies of ecology. It attempts to summarize provisionally various observations, from scattered outcrops and wells, on the stratigraphic relationships of some reefs in the Middle East.

The principal types are fringing-reefs, barrier-reefs, shoal-reefs, reef-knolls, and bank-reefs, the last term being used to indicate reef-complexes formed over submerged highs, generally of tectonic origin. It is shown that, under ideal conditions, the back-reef, reef, and fore-reef (basinward) facies-zones of a reef-complex can be differentiated by petrographic and micropaleontologic criteria recognizable in thin sections, even of well cuttings; and that by this means it is commonly possible to infer the position and history of a reef from a series of local observations.

The term "reef-complex" is applied to the aggregate of reef limestones and the calcareous rocks genetically (?) associated with them.

The zone of interdigitation between a reef-complex and the basinward sediments is considered to be particularly favorable for the generation, migration, and accumulation of oil, subject to certain conditioning factors such as the relative development of primary and secondary porosity; or of primary or secondary lime cementation in fore-reef limestones which, if porous, may act as carrier beds and reservoirs.

It is shown that the reefs grew in certain sub-littoral zones and on some submerged highs of epeiric seas covering the Arabian foreland, in which deposition of the globigerinal chalk-marl facies predominated.

INTRODUCTION

Interest in the reef limestones of the Middle East arose through the discovery that the rich, productive formation of the Kirkuk oil field is a reef-complex of Tertiary age, while two other fossil reefs respectively of Lower-Middle and of Upper Cretaceous age are highly bituminous at their outcrops.

Examination of these and other occurrences led to the conclusion that some detrital limestones associated with the reefs themselves are more important than the latter both in relative bulk[3] and in their rôle as oil carrier beds and reservoirs.

It is obvious that reef research by petroleum geologists should include not

[1] Read by D. A. Greig, before the Association at St. Louis, March 16, 1949. Manuscript received, March 31, 1949.

[2] Research geologist, Iraq Petroleum and Associated Companies, Geological Research Centre, London, England. This paper is published by permission of the management and chief geologist of Iraq Petroleum and Associated Companies, Limited. It repeats and expands the substance of a short address given to Section E of the XVIIIth International Geological Congress, London, 1948. The manuscript was prepared to be in time for the Reef Symposium meeting at St. Louis (March, 1949) and was most ably condensed and presented on the writer's behalf by D. A. Greig. Subsequently the writer took opportunities to visit or re-visit critical sections in the Middle East, and as a result the original text has been reworded in parts without requiring any change of substance as summarized for the St. Louis meeting.

In developing his ideas, for which he accepts sole responsibility, the writer has drawn upon the unpublished work of colleagues too numerous to mention individually. However, reference should be made to the contributions of W. Kitchin and D. A. Greig who participated in the writer's early studies of the Kirkuk reef limestone. Discussions with R. G. S. Hudson, who is well acquainted with reef-building conditions in the Carboniferous of England, have been particularly helpful. A. H. Smout and G. F. Elliott have given valuable assistance in the preparation of micrographs and manuscript, respectively. To all these collaborators the writer offers his sincere thanks.

[3] W. H. Twenhofel, *Treatise on Sedimentation* (1926), p. 228. The Williams and Wilkins Company, Baltimore.

only the reefs *sensu stricto*, but all genetically(?) associated sediments which may play their parts in oil history.

Any aggregate of this nature is here termed a *reef-complex*.

Examples of the latter are numerous at many horizons in the Middle East, and some individual complexes probably extend over thousands of square miles; however, it is not the object to give here any descriptive catalog of them which would be impossible in the present short paper. The latter is intended only to offer for discussion in the reef symposium some general observations and hypotheses which are provisional in nature, since investigation of reefs in the Middle East has never been developed systematically as a research project.

Nevertheless, a considerable volume of information has been accumulated incidentally, over a period of more than 20 years, from scattered outcrops and well sections.

In addition there have been brief opportunities for studying accessible parts of recent reefs in Caribbean, Persian Gulf, and Levantine waters.

By collating the data obtained, it becomes apparent that there are certain common features in reef architecture, expressed in the detailed composition, distribution, and environmental relationships of facies-zones, which led to the concept of ideal reef-complexes of various types.

Although departures from the ideal are numerous in nature, due to the interplay of many independent variables, they are more easily comprehended if examined against the background of the ideal pattern.

The extension of reef research to all genetically(?) associated sediments, which are mainly hard limestones, introduces questions of technique. Destructive sedimentary analysis obliterates important features of the rocks, while not all delicate petrographic and micropaleontologic distinctions, which are necessary for this study, are sufficiently apparent when specimens are viewed in the rough under a binocular microscope.

Critical details of limestones can be appreciated best by examination of thin sections[4] which, to the eye of an experienced observer, tell a graphic story of the environment in which the rocks were deposited.

Great stress is naturally laid on the ecologic[5] and correlation value of foraminifera and other microfossils, many of which can be recognized, at least generically, in sections.

[4] Sections which are not required for mineralogical studies need not be cut to standard thicknesses, and if all unnecessary refinements, even cover slips, are eliminated, a good operator with mechanical equipment can make 20 sections or more per day.

Even small drill cuttings yield valuable information by this method.

[5] E. H. Myers, "Ecologic Relationships of Some Recent and Fossil Foraminifera," *Rept. Comm. Marine Ecology as Related to Paleontology, 1941–42, Natl. Res. Council* (Washington, 1942), pp. 31–36.

———, "Ecologic Relationships of Larger Foraminifera," *ibid., 1942–43* (1943), pp. 26–30.

R. D. Norton, "Ecologic Relations of Some Foraminifera," *Bull. Scripps Inst. Oceanography*, Tech. Ser., Vol. 2, No. 9 (1930), pp. 331–88, text-figs. 1–3.

F. R. S. Henson, *Middle Eastern Tertiary Peneroplidae (Foraminifera), with Remarks on the Phylogeny and Taxonomy of the Family*, 8 vo, 70 pp., 10 pls., 3 text-figs. West Yorkshire Printing Company, Wakefield, England (1950).

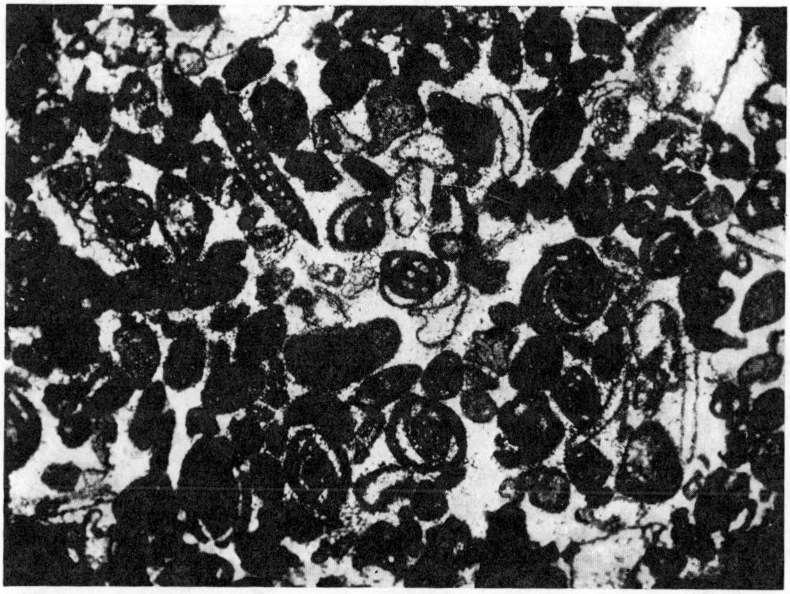

FIG. 1.—Back-reef shoal limestone; medium grade. ×12; Kirkuk well; upper Oligocene.

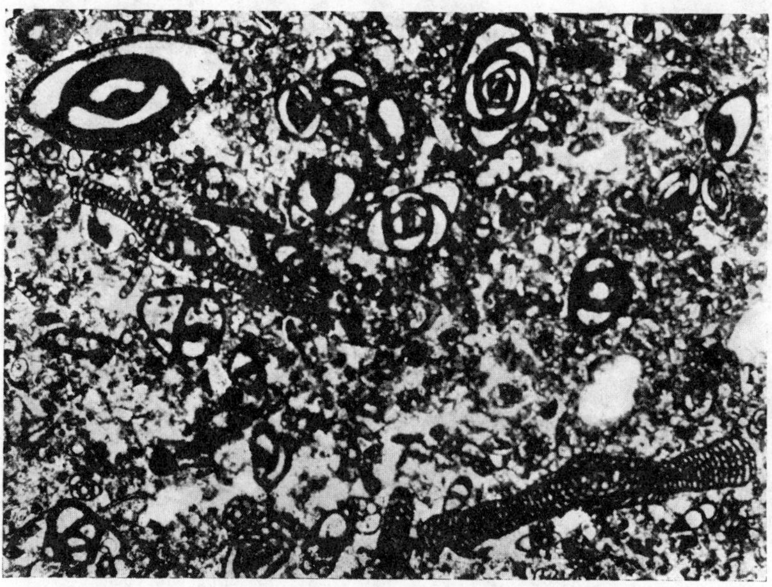

FIG. 2.—Back-reef shoal limestone, fine grade. ×15; Kirkuk well 19; upper Oligocene.

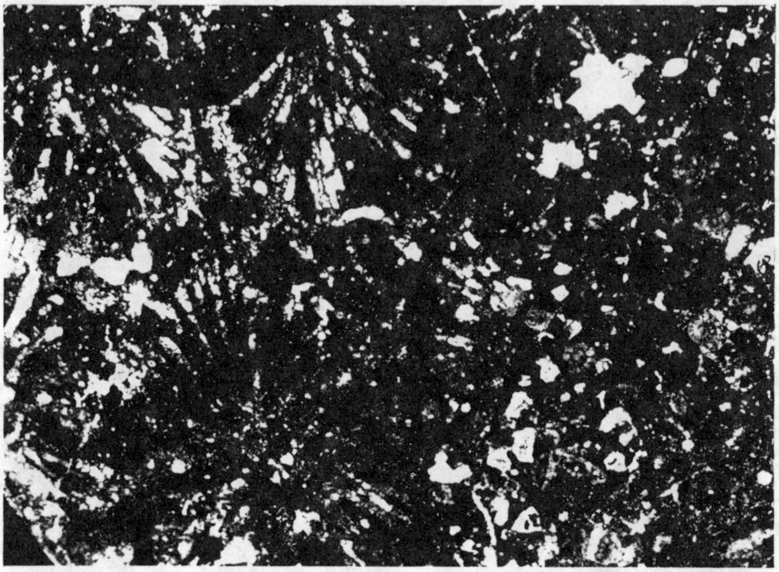

FIG. 3.—Reef limestone. ×7.5; Kirkuk well 27; upper Oligocene.

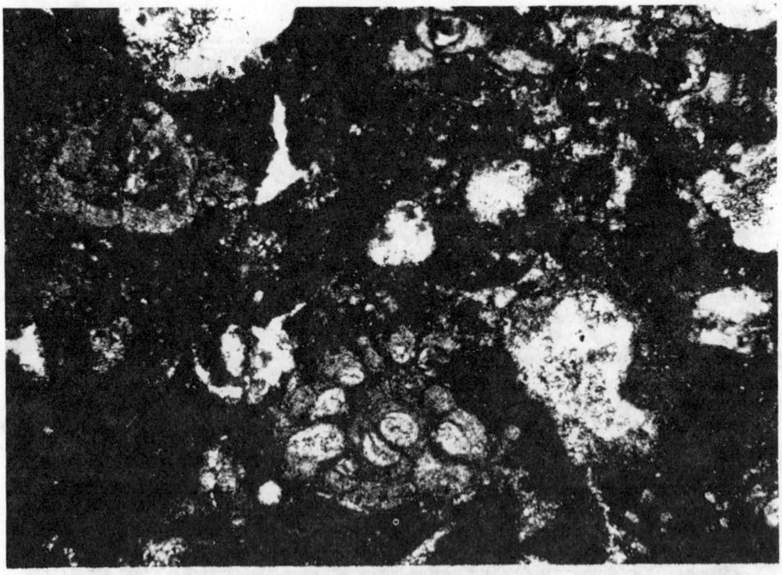

FIG. 4.—Reef limestone. ×30; Kirkuk well 27; upper Oligocene.

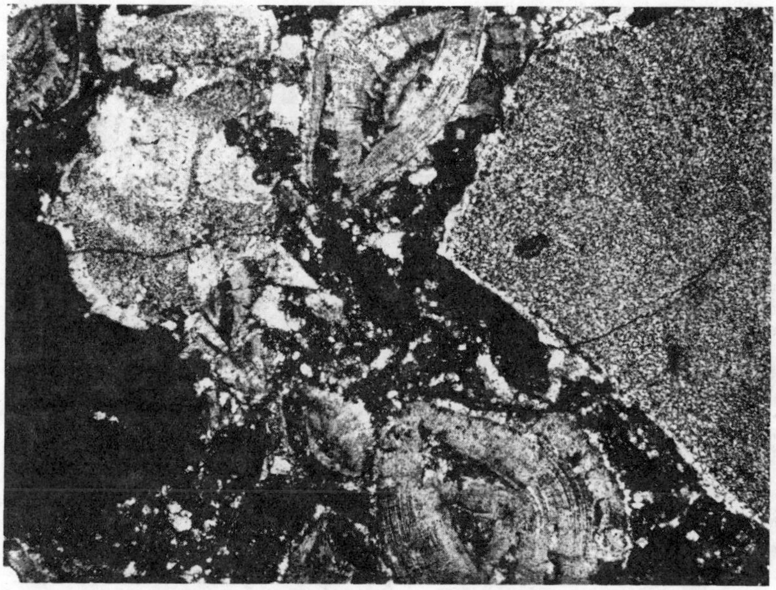

FIG. 5.—Reef talus, fine grade; or fore-reef shoal limestone, coarse grade. ×30; Kirkuk well 19; Oligocene.

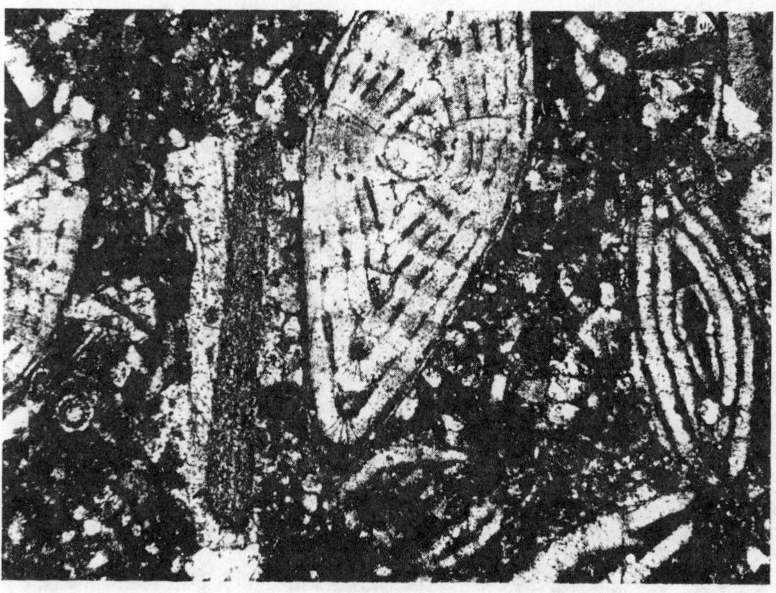

FIG. 6.—Fore-reef shoal limestone, medium grade. ×30; Kirkuk well 25; lower or middle Oligocene.

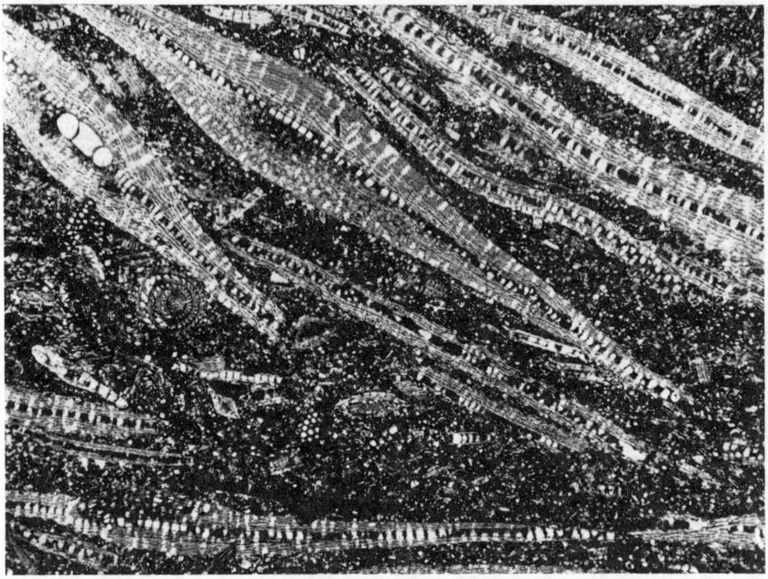

Fig. 7.—Fore-reef shoal limestone, fine grade. ×7.5; Kirkuk well 27; Oligocene.

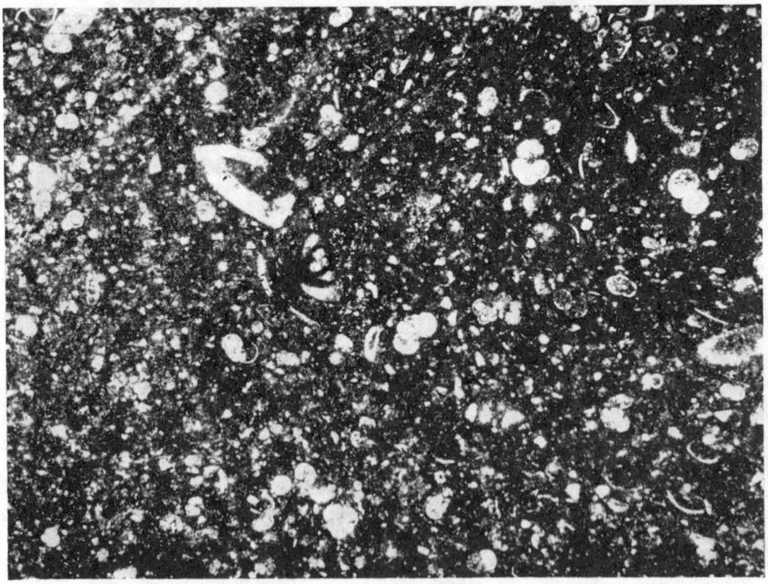

Fig. 8.—Basinal limestone with traces of calcareous detritus. ×30; Jebel Abd El Aziz, Syria; Oligocene.

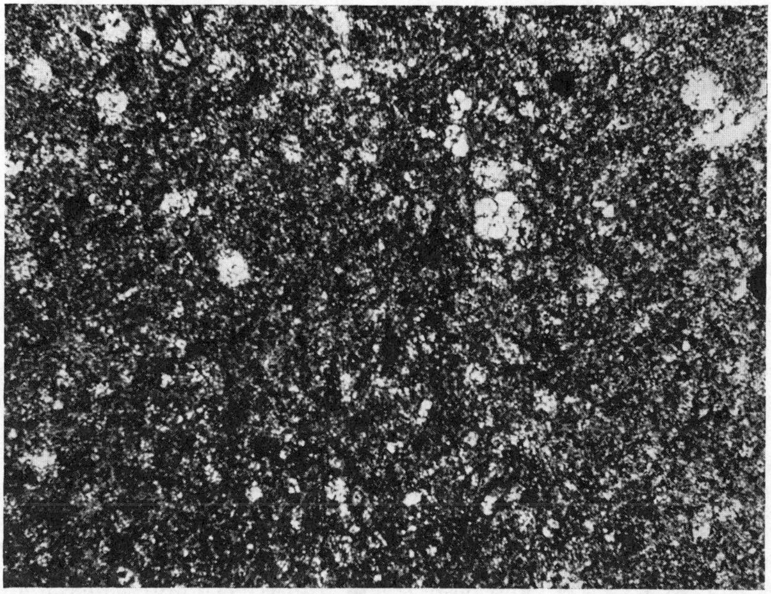

FIG. 9.—Basinal chalk. ×30; Jebel Hayane, Syria; upper Eocene.

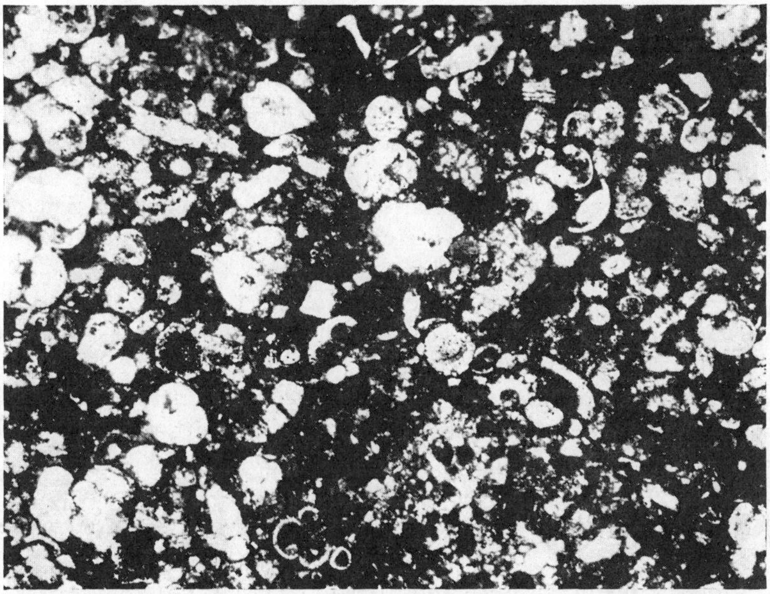

FIG. 10.—Glauconite-microfossil concentrate at zone of stratigraphic condensation. ×30; Kirkuk well 14; Eocene.

The writer's results rest largely on orthodox thin-section studies, covering thousands of samples of reef material, co-ordinated with evidence of field relationships. The picture presented is believed to be well founded, and indeed it involves little that is new; but it may be over-generalized, and much planned investigation of details is still needed.

In the following discussion, some deposits are described (reef-patches and open shoals—see Table I) which may be considered outside the subject of the reef symposium. They are, however, borderline cases of sediments genetically (?) associated with reefs and are included here for comparative purposes.

GENERAL STRATIGRAPHY

That region of the Middle East which is considered in this paper lies south of the Turkish, and west of the Iranian, frontiers, but excludes Egypt and western and southern Arabia. It is part of the Arabian craton which has subsided slowly and intermittently beneath epeiric seas.

Periodic uplifts of an epeirogenic character have produced unconformities in the sedimentary column, and other observed complications suggest the effects of sharp, vertical movements probably due to deep-seated block faulting[6] along NW.-SE., NE.-SW., E.-W., and N.-S. trends. But true compressional folding was more restricted.

In the northwest there are indications of Caledonian orogeny; and various authors have concluded that there was important thrusting associated with the Austrian and successive Alpine movements in the Tauric arc of Suess, and its continuation in the Zagros and Oman ranges. However, in the region with which we are concerned, there is no definite evidence of compressive folding before late Miocene-Pliocene time when the present pattern of anticlines and synclines was produced, and was probably conditioned to some extent by the pre-existing block structure.

Until the Mio-Pliocene folding, the greater part of the region remained relatively stable, but was affected to the extent that orogenies in adjacent areas stimulated contemporaneous vertical movements.

Our story begins at the close of the Jurassic when extensive uplift led to the emergence of some areas while marine sedimentation continued elsewhere in open sea or in residual morpho-tectonic basins.

Some of these accumulated chemical deposits (anhydrite *et cetera*), while in other areas marine limestone and shale sedimentation, with marginal sands, was uninterrupted from the Jurassic into the Cretaceous.

Marine transgression began again within the Lower Cretaceous interval and was widely extended in the Cenomanian when most of the Arabian shelf was submerged.

Thereafter the history was complicated by the effects of successive Alpine

[6] The importance of these movements was first emphasized in detail by E. J. Daniel in unpublished work.

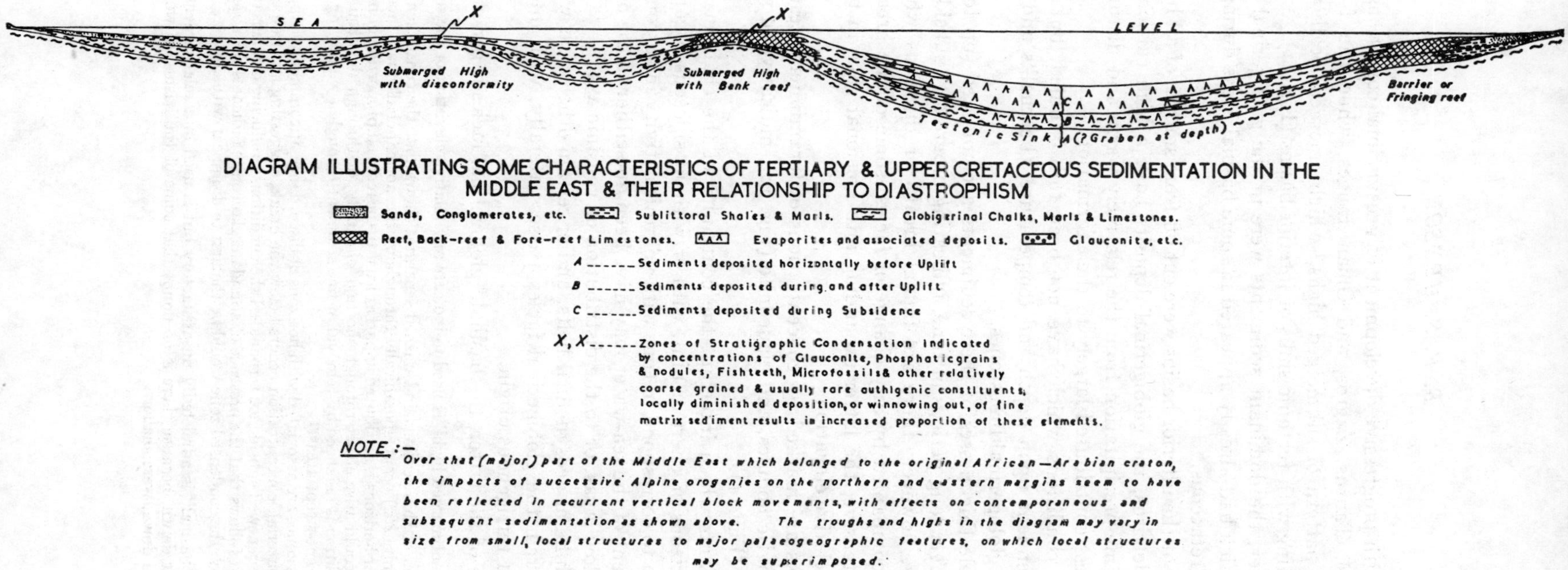

FIG. 11.—Characteristics of Tertiary and Upper Cretaceous sedimentation.

movements, with progressive development of geosynclinal conditions in the Kyrenia (Cyprus), Taurus, Zagros, and Oman ranges, while the foreland was divided into a pattern of submerged highs,[7] which emerged locally and temporarily, alternating with tectonic sinks or interior basins (Fig. 11).

In the latter, the dominant sediments were in the globigerinal chalk-marl-shale facies which had already appeared in some basinal areas during the Lower and Middle Cretaceous.

Since most of the interior basins were cut off from sources of all but the finest land-derived detritus, the globigerinal deposits have a high carbonate content and primary lime precipitation from the relatively warm and disturbed surface waters doubtless contributed largely to their formation.

In certain conditions which have not been fully studied, but which were doubtless dependent upon depth and temperature, the chalks and marls have hardened into globigerinal limestones.

Considerable thicknesses of these sediments accumulated (up to 4,500 feet) continuously over long periods of time, indicating progressive subsidence.

Frequently, however, there are faunal breaks or diastems which are not necessarily accompanied by any change of facies; some of the breaks are local and others extensive, the latter expanding into erosional gaps in the sequence over areas of relative elevation.

Such breaks are marked by concentrations of foraminiferal tests, fish fragments, phosphatic nodules, and glauconite (Fig. 10), indicative of stratigraphic condensation[8] (Fig. 11).

Other variants which interrupt the continuity of the chalk-marl facies are occasional horizons of chert or flint, with or without siliceous organisms, which generally seem to correspond in time with volcanic activity in surrounding areas; silica enrichment of the sea-water may have been due either to the outpouring of submarine pillow lavas, or to the distribution of volcanic ash.

In the peripheral geosynclinal belts, and in certain other areas where marked faulting occurred, the globigerinal facies passes laterally, with interdigitation, into clastics of terrigenous origin.

In absence of the latter, the basinal,[9] globigerinal facies is generally replaced

[7] The term "submerged high" is used to indicate any relative elevation of the sea-floor.

[8] Examination of thin sections and washed separates shows that the constituents of these concentrates (glauconite *etc.*) occur sporadically throughout the normal chalk-marl facies. Their abundance at particular horizons is attributed in part at least to the reduced proportion of matrix sediment; and this in turn could arise either through shortage of supply, or through the attainment of a base level where deposition is balanced by removal of fine sediment through a winnowing action of water in shallow or current-swept areas.

If glauconitic zones are associated with horizons of discordance, it commonly happens that, apart from reworked material, concentration occurs below the break, suggesting a period of stability after uplift, and before renewed subsidence terminated the conditions producing condensation.

In this case, it follows that the presence of abundant glauconite indicates the absence of erosional unconformity since the condensed zone would be the first to disappear with erosion.

[9] The term "basinal" is admittedly unsatisfactory but is used in a relative sense to avoid any implication about depth, because there is no doubt that some of the basins, in which globigerinal deposits were laid down, were shallow.

toward shore lines or submerged highs by sheet-like shelly and foraminiferal limestones which are phosphatic and siliceous in some environments; or into reef-complexes which interdigitate down-grade with the basinal facies (Fig. 11).

The development of these reef-complexes is obviously conditioned mainly by depth and temperature, but there are other factors which are not so clear.

In many cases the globigerinal facies persists both above and below breaks marked by discordance or stratigraphic condensation, or by evidence of disturbed, shallow-water sedimentation.

In other cases, where the environment seems to have been similar, reef-complexes developed; yet again the latter may follow the globigerinal facies where no break in sedimentation is apparent, though this observation is admittedly dependent on the accidents of exposure.

It is reasonably supposed that coral larvae require some clean, hard surface for attachment before a colony can develop; yet there are reefs in the Middle East which rest on globigerinal marl.

It seems likely that various deposits, such as lenticles of pebbles, phosphatic nodules, or molluscan shells, might provide the starting points for coral reefs which, in favorable environments, would furnish thereafter their own foundations of talus for progressive expansion.

However, the writer believes that many coral reefs are based on patches of algae or bryozoans which can develop even among soft sediments.

REEF-COMPLEXES OF MIDDLE EAST

CLASSIFICATION

Descriptive names currently used for reefs of various types are not entirely satisfactory as a basis of classification because the concepts of reef-form, reef-origin, and reef-environment are somewhat confused in them.

No attempt at a revised classification can be made in this short paper, but some indication must be given of the sense in which the various common terms are used here, with brief references to examples in the Middle East.

Fringing-, *barrier-*, and *bank-reefs* are regarded essentially as environmental categories.

Differentiation of fringing- and barrier-reefs is not everywhere easy when relationships are in large part concealed by later deposits. The presence or absence of a definite lagoon between reef and shore, as distinct from shallow shoal waters, is accepted as the criterion for a barrier reef.

Thin, patchy, fringing-reefs of Maestrichtian, lower Eocene, middle Eocene, upper Eocene, Oligocene, and Miocene ages, are spread over the foreland shelf of the Iraq-Iran geosyncline; they are known also in the folded Zagros ranges at the east, and are there conditioned to some extent by morpho-tectonic features due to the Alpine movements.

The Kirkuk productive limestone is part of a compound fringing-reef belt at least 250 miles in length, ranging in age from middle Eocene to Miocene; however,

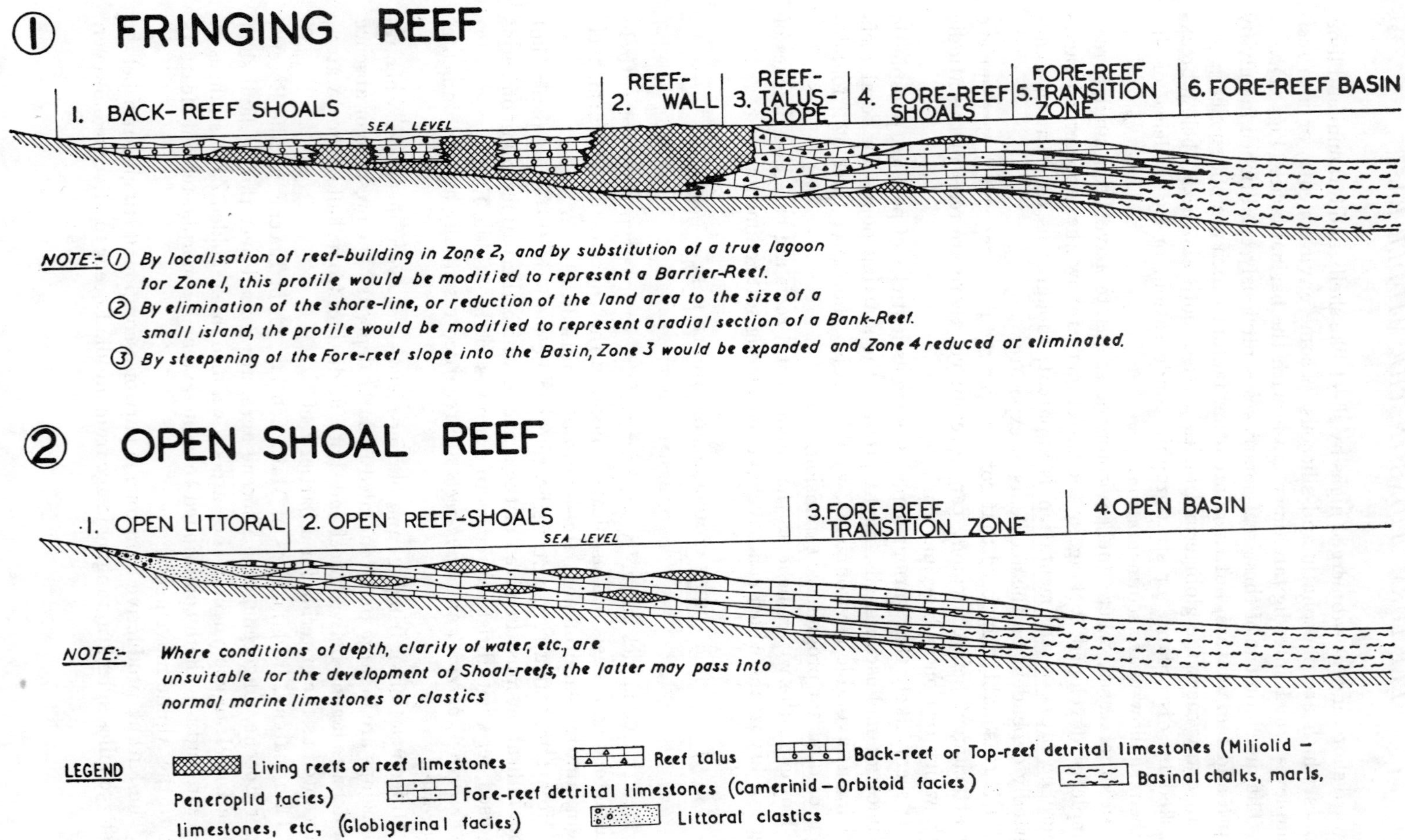

FIG. 12.—Reef profiles illustrating sedimentary environments.

the Eocene reefs are of barrier type in the Kirkuk-Mosul area, passing northeastward into chemical limestones of an extensive back-reef lagoon.

Miocene fringing-reefs are widely developed in northwest Syria and Lebanon.

Bank-reefs.[10]—This term is extended to include any large reef-growths, generally irregular in shape, which developed over submerged highs of tectonic or other origin, and which are more or less completely surrounded by deeper water.

Bank-reefs of Maestrichtian-Campanian age occur in northern Iraq, and the Paleocene-lower Eocene reef-complex of Jabal Sinjar may be of the same type. They are common in the lower Eocene and middle Eocene of western Syria, but the best examples are in the Oligocene of the Palmyra area (Jabal Abiad *et cetera*).

Reef-terms which imply form rather than environment are as follows.

Reef-knolls are small growths which developed locally and upward rather than outward. Although it has been claimed that the original reef-knolls of Tiddeman[11] are erosional features and not local reef-growths, there is no reason why this useful descriptive term should not be retained in accordance with Tiddeman's first concept, where the latter does apply. Examples occur in the Middle East but are of no importance for purposes of the present paper.

Atolls.—True atolls are not known in the Middle East.

In addition to these familiar form-categories of reefs, the writer finds the following useful.

Reef-wall.—A wall-like up-growth of dead and living coral and/or other reef-building organisms (Fig. 12, fringing-reef profile, and Table I) reaching intertidal level and serving therefore as a partial barrier between adjacent environments.

Wall-reefs may occur as parts of fringing-, barrier-, or bank-reefs, and, in circular form, are an essential feature of atolls.

Shoal-reefs.—Some authorities use this term for bank-reefs, but the writer prefers to apply it to all formations in which reef-growth develops in irregular patches amidst submerged shoals of calcareous débris (Fig. 12 and Table I).

Algal and bryozoan reef-patches are most common, but corals also occur, particularly the small, deeper-water species in submerged shoals.

If the bioherms are small and thin, they do not conform with the accepted definition of reefs, but, as the differences are those of degree only, they are discussed here as borderline cases.

Shoal-reefs may compose parts, or the whole, of fringing-, barrier-, or bank-reefs, and they may develop locally and sporadically under suitable, clear-water conditions, in areas where other types of sediment predominate. It is suggested

[10] W. M. Davis, "The Coral Reef Problem," *Amer. Geogr. Soc. Spec. Pub.* 9 (1928), p. 19.

[11] R. H. Tiddeman, "On Concurrent Faulting and Deposit in Carboniferous Times in Craven, Yorkshire, with a Note on Carboniferous Reefs," *Rept. Brit. Assoc.*, Newcastle-upon-Tyne, 1889 (1890), pp. 600–03.

———, "On the Formation of Reef-Knolls," *Rept. Brit. Assoc.*, Bradford, 1900 (1900), pp. 740–42.

that the seaward extension of the submerged platform, on which true coral reefs are commonly found to rest,[12] is generally accentuated by, and in some cases may be entirely composed of, fore-reef shoal material (Fig. 12, fringing-reef profile).

In the Cretaceous and Tertiary of the Middle East there are numerous patchy occurrences of algal and foraminiferal limestones, with some corals, and with associated detrital limestones, which do not seem to be connected with true coralline fringing-, barrier-, or bank-reefs; these are interpreted provisionally as open shoal-reefs or reef-shoals, recent examples of which can be seen in parts of the Persian Gulf.

Judged by beach material and other indications, submerged algal shoal-reefs are probably in process of development off-shore at Tripoli, Lebanon.

Rudist shoal-reefs are common locally in the Lower-Middle Cretaceous limestone unit which covers an enormous area of the Middle East, attaining in some places a thickness of several thousand feet. This limestone shows a fairly constant range of lithological components which include (1) well bedded, dense and porcellaneous, chemical(?) limestones with few fossils, (2) massive, coarse or fine, crystalline, calcitic, or dolomitic limestones, probably due to alteration, (3) limestones, micro-detrital in part, with *Orbitolina et cetera*, (4) limestones, micro-detrital in part, with *Dicyclina*, *Praealveolina* spp., Miliolidae, Peneroplidae, *et cetera*, (5) rudist reefs, (6) oyster beds and other shelly layers, and (7) subordinate marls and anhydrite lenticles.

In some areas, massive rudist reefs increase to make up an important part of the formation, and are associated with detrital, foraminiferal limestones; it is probably correct to classify these developments as bank-shoal-reefs.

The classification of reefs adopted here is convenient for many purposes, but it is obvious that the various categories overlap considerably under natural conditions.

Thus a reef-knoll, if extended laterally to large dimensions, would become a fringing-, barrier-, or bank-reef, depending on its physiographic relationships.

A barrier-reef may merge into a fringing-reef if the shoreward lagoon develops any considerable amount of reef-growth; and fringing-reefs may pass long-shore into deeper-water shoal-reefs.

Parts of bank-reefs may have the character of barrier-reefs or fringing-reefs in relation to large islands emerging in a bank.

These intergradations should cause little confusion in the consideration of recent reefs where each case may be viewed as a whole, but large parts of fossil reefs are necessarily obscured and classification may be more difficult.

[12] T. W. Vaughan, "Corals and the Formation of Coral Reefs," *Ann. Rept. Smithsonian Inst. 1917* (1919), pp. 228–33.
H. T. Stearns, "An Integration of Coral Reef Hypotheses," *Amer. Jour. Sci.*, Vol. 244 (1946), p. 247 (note 10).

FACIES-ZONES IN REEF-COMPLEXES OF MIDDLE EAST

Figures 12 and 13, and Table I, illustrate in self-explanatory fashion the writer's concepts of ideal reef-complexes, based on the collation of fragmentary information obtained in the field and in the laboratory.

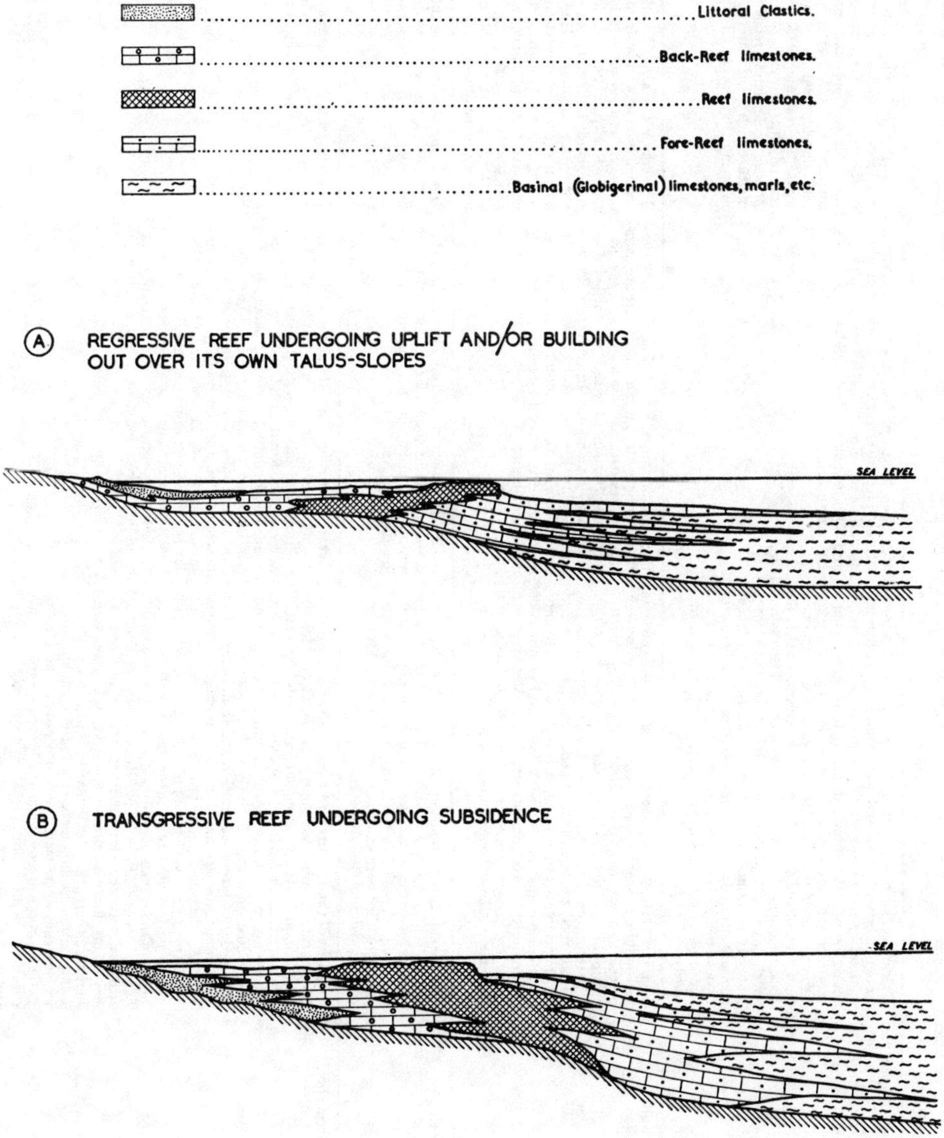

FIG. 13.—Migration of reefs.

TABLE I

CLASSIFICATION AND DESCRIPTION OF FACIES-ZONES IN REEF-COMPLEXES OF MIDDLE EAST

Classification	*Sedimentation*	*Indigenous Fauna*	*Remarks*
Ref. Fig. 12 (Fringing-reef profile) I(1) Back-reef shoals Figs. 1, 2	Calcareous sand and mud composed mainly of comminuted débris of reefs and indigenous organisms *well rounded by surf-erosion;* cemented by rapid lime-precipitation in warm, shallow water, to form hard, poorly stratified *Miliola* limestones of porcellaneous texture and sub-conchoidal fracture. In some varieties, the proportion of fossil and detrital elements is so reduced that the rock is almost pure chemical limestone. These shoal rocks may interdigitate with littoral clastics but are typical only if formed in clear water. Reef-patches occur sporadically and may be contemporaneous local growths or dead residuals of former reef-walls, eroded by surf and over-ridden by shoals as the living reef builds out seaward	Miliolidae, Peneroplidae (large and small) and Alveolinidae are characteristic, with some Planorbulinidae, Rotaliidae, Amphisteginidae, and Orbitolinidae. Reef-builders, mollusks, etc. are sporadic	Typical *Miliola* limestones occur in shallows of barrier-reef lagoons and in littoral shoal areas of fringing-reefs. Over bank-reefs they form top-reef shoals. Along flat, foreland shores and in analogous shallow-water environments where reef development is in patches rather than belts, the reef and back-reef facies may be intricately intermingled. *Miliola* limestones, recognizably modified by mixture with other material, may also be formed in open littoral zones (II, I)
I(2) Reef-wall Figs. 3, 4	Vigorous vertical, and in some places lateral, growth of reef-building organisms with cavities filled in by reef-breccia and by rapid bio-chemical and physico-chemical precipitation of lime. In many places volume proportion of matrix greatly exceeds that of organic structures. Resulting rock mass is unstratified, very irregular and lithologically heterogeneous, with dense, commonly porcellaneous matrix, but with zones and channels of high porosity due to coral structures etc. Algal reefs are much less porous in this respect and do not as a rule form reef-walls without supporting corals	Reef-building organisms (corals, algae, bryozoans, rudists, etc.); foraminifera rare and generally restricted to sessile forms such as *Rupertia* and *Homotrema* with thick-shelled *Amphistegina*, *Rotalia*, etc. In relatively still water, or below breaker-level on fore-reef side, other foraminifera characteristic of zone 1 (4) may abound at the reef-margins, probably living in crevices etc.	The concept of a reef "wall" is based on its observed effect of maintaining a sharp line of division between back-reef and fore-reef facies-zones. Reef-patches which do not have such effect are excluded from this category. Reef-wall has not been recognized as such in fossil form probably because it is eroded and over-ridden by later transgressions. Moreover, it is narrow compared with its contemporaneous back-reef and fore-reef zones and appears in fossil reefs merely as seaward limit of irregular reef zone over-ridden by shoal deposits
I(3) Talus-slope Fig. 5	Cross-bedded accumulation of unsorted reef-breccia broken off by breaker-erosion and precipitated below breaker-level; *fragments predominantly angular*, and diminishing markedly in size away from reef. Interstices generally filled with calcareous mud; primary lime precipitation diminishes with depth	Some deeper-water reef-building organisms grow upon talus-slopes. Echinoidea and large foraminifera as in zone 1 (4) are abundant below breaker-level	Development of this zone depends largely on the physiography of reef-wall and seaward slope. Coarse reef talus may be confused with reef-rock *in situ* in samples from wells
I(4) Fore-reef shoals Figs. 6, 7	Deposits of this zone are formed generally below breaker-level, and are subjected mainly to effects of tidal ebb and flow with little vigorous erosion. Undertow carries some material out from reef-front and deposits it in bare or weed-covered shoals of well sorted calcareous muds and sands; these are relatively stable because the elements, derived from reef-erosion and from decay of organisms *in situ*, are angular; their average grade-size generally decreases basinward. Some calcareous sands are formed largely of comminuted débris of echinoids and large foraminifera. Coarser detritus from the remains of indigenous organisms is swept by currents into banks, bars or lenticles which are preserved as ill sorted fossil microbreccia. Reef-patches, mainly of algae and bryozoans, but with some deeper-water(?) corals, form, here and there, thin, lenticular beds of reef limestones. Cementation by lime precipitation is common in the reef-patches where biochemical processes are active, but variable in calcareous sands and breccias, probably according to depth conditions. Many fore-reef shoal limestones retain high, primary porosity	*Large* foraminifera are abundant and robust including, according to age, Nummulitidae, Orbitoididae, Miogypsinidae, Peneroplidae, Alveolinidae, Rotaliidae, Orbitolinidae, Lituolidae. Local concentrations of particular species of large foraminifera, e.g., *Nummulites* spp. and *Lepidocyclina* spp., suggest fossilization of communities *in situ* (Fig. 7). Fragmentary remains of Echinoidea are abundant. Reef-building organisms and Mollusca are sporadic. There are some indications that Orbitoididae may occur over wider range of depth conditions than most other large foraminifera	Deposits of this zone may be difficult to differentiate from those of zone II (2) excepting by the mixture of allocthigenous material

TABLE I—(*continued*)

Classification	*Sedimentation*	*Indigenous Fauna*	*Remarks*
I(5) Fore-reef transition zone	Tongues of fore-reef detrital limestones alternating and intergrading with sediments of zone I (6)	Mixed faunas of zones I (4) and (6)	This zone is probably the broadest in most reef-complexes
I(6) Fore-reef basin Figs. 8, 9	Steady deposition of fine-grained, globigerinal chalks, shales, and limestones, or exceptionally of evaporites	Globigerinidae and associated small, pelagic foraminifera, etc.	This zone lies outside limits of reef-complex but is listed to complete picture of lateral variation
Ref. Fig. 12 (Open-shoal reef profile) II(1) Open littoral	Terrigenous clastics, detrital limestones, oölites, shell breccias, etc. Characteristic of this environment is mixture of indigenous and allocthigenous detrital elements, ordinarily rounded; and greater variety of deposits compared with those of zone I(1). Miliolid-peneroplid limestones occur, but are bedded, impure, and petrographically heterogeneous	Fauna very varied; Peneroplidae, Miliolidae, Alveolinidae, Amphisteginidae, Rotaliidae, and Elphidiidae are locally abundant and mixed with pelagic foraminifera, and with those of zone II(2). Mollusca, Echinoidea, etc. form shell-banks	Limits between this zone and II(2) are indefinite
II(2) Open-reef shoals	As in I(4), but with greater mixture of allocthigenous material. Reef-patches may be common and lithology generally is more varied, but some limestones may be undifferentiated from those of zone I(4)	As in I(4), but with admixture of littoral organisms (Miliolidae, Peneroplidae, etc.) from zone II(1) carried out by under-tow or by floating weeds etc.	Open shoals and reef-patches probably occur in clear-water areas of such depths that only algae, bryozoans and less vigorous deeper-water corals can exist, so that no intertidal reef-walls are developed. This zone may pass laterally into normal limestones or clastics
II(3) Fore-reef transition zone	Alternating, and intergrading deposits of zones II(2) and (4)	Mixed faunas of zones II(2) and (4)	Probably most extensive zone
II(4) Open basinal	As in I(6)	As in I(6)	See I(6)

Notes.—

1. The idealized arrangement of depositional zones as illustrated is subject under natural conditions to many irregularities and intergradations which are more readily visualized than described. Owing to storm-wash, gaps in reef-wells, etc., the faunas of otherwise sharply separated zones may be subject to some degree of mixture, but the dominant indigenous faunas are nearly everywhere recognizable as such. Doubt may arise in the case of foraminifera such as the Alveolinidae which are widely distributed, due either to varied habitat or to their form favoring transport by currents etc.

2. Besides the characteristic sediments described, there are other types which appear in reef-complexes, but are too restricted and diverse to merit attention here. Back-reef evaporites and aragonite mud are not known in the Middle East, but "chemical" limestones deposited in a barrier-reef lagoon occur in one area. Back-reef oölites are rare, but oölitic limestones occur in zone II(1).

3. Equivalent facies of different ages have in common the micro-characters described, but may be recognized by fossil species appropriate to their ages, and commonly by minor but constant petrographical differences which serve as useful empirical criteria for observers experienced in the region.

The question may well be raised about how far we are justified in regarding the detrital, shoal limestones described in Table I, apart from the obvious case of reef-talus, as sediments "genetically associated" with reefs.

The back-reef, miliolid-peneroplid limestones might be classified as normal, littoral deposits capable of occurring in areas where there are no reefs at all; and the fore-reef and open-shoal types might develop in similar conditions as normal sublittoral limestones.

It is, in fact, true that recognizable reef-débris is commonly a very minor constituent of all these detrital limestones and may be totally absent.

Genetic association with reefs can not, therefore, be claimed on the ground that reef material is an essential component of the detrital limestones.

Nor is it likely that there is any biological interdependence between the faunas of the reefs themselves and those of associated shoal deposits, in which comminuted indigenous fossil remains are the most important elements.

The Miliolidae and Peneroplidae of the back-reef zone, and the Nummulitidae, Orbitoididae, *et cetera*, of the fore-reef and open-shoal zones, occur also in other sediments unconnected with reefs, for example, in terrigenous clastics.

But the facies-zones of Table I are differentiated by the aggregate characters of the rocks, expressing specialized, clear, shallow-water conditions of deposition with rapid building and collateral fragmentation of calcareous structures; and these conditions, according to the writer's observations, are commonly found in conjunction with developments of reef-bioherms.

It remains for future research to show whether the reefs themselves are essential, or incidental, components; or whether the distinction is irrelevant because the environment alone is the genetical common factor.

Meanwhile, the writer's concepts of ideal reef-complexes have only an empirical validity based on frequently observed relationships; but, if only on this basis, they have useful applications in stratigraphic synthesis in the Middle East, subject to the possibility of exceptions.

Complete horizontal sequences of contemporaneous facies-zones (Fig. 12) have never been observed at outcrop, but may be inferred with reasonable certainty from surface and subsurface correlations.

No contradictory evidence has been obtained, and when two facies-zones are found to interdigitate or to intermingle in one section, which is not tectonically or otherwise complicated, they are without exception those which are adjacent in the ideal sequence.

Vertical successions through migrating reef-complexes (Fig. 13) naturally vary according to their locations with respect to the initial and final positions of the reef, and according to the rate and degree of uplift or submergence.

Rapid drowning of a reef, for example, may result in the direct superposition of basinal over reef or back-reef deposits.

Only one, or several, facies-zones may be represented at a locality, but sections which show a complete vertical succession are rare, because it is only in a very restricted belt of a migrating reef-complex that such coincidental overlap can occur; and in most places it does not occur at all.

Nevertheless, perfect examples are available in the Middle East, as in the productive formation of the Kirkuk oil field (Baba Gurgur dome) which reveals a succession from globigerinal limestones at the base, through fore-reef, then reef-limestones, to back-reef *Miliola* limestones which are partly eroded at an unconformity.

Paleontological evidence indicates, within the limits of the fossil record, that each facies-zone cuts upward across time-planes in a basinward direction, this history ranging from middle Eocene to Miocene with some complications due to intervening movements, *et cetera*.

This is a case of a regressive[13] reef moving basinward through uplift which culminated in emergence (Fig. 13A).

Transgressive[13] reefs, which were drowned by submergence and burial under basinal deposits, are common in the region (Fig. 13B).

There are also examples of complete sedimentary cycles being expressed in the sequence of reef and basinal facies.

Thus, a Lutetian-upper Eocene limestone unit in the Zagros ranges began, in suitable environments, with a transgressive back-reef and reef-facies which was succeeded vertically first by Lutetian (nummulitic) fore-reef limestones, and then by globigerinal limestones; continuing upward without a break, the nummulitic fore-reef facies returned with upper Eocene fossils, and was covered by a regressive reef and back-reef zone of late Eocene age. Each change is accompanied by alternations of facies and it is rarely that the full sequence is seen in a single locality, but the story is clear from the evidence of offset sections.

In the Kirkuk area (Khurmala and Avanah domes) the Lutetian reef-limestone was probably developed as a barrier-reef, because contemporaneous deposits at the northeast are well bedded chemical limestones with a very meager stunted fauna, suggesting deposition in an extensive back-reef lagoon under rather stagnant conditions. These limestones pinch out farther northeast into littoral conglomerates.

In both vertical and horizontal sequences, some facies changes are remarkably sharp, as at Kirkuk where reef conditions must have approximated the ideal shown in Figure 12, fringing-reef profile; but generally they are gradual and accompanied by interdigitation and mixture of adjacent facies-zones, as in the Lutetian-upper Eocene limestones, which may confuse local sections while fitting consistently into the picture of the complex as a whole.

Over broad banks, the classification of reef-zones loses its significance and limestones of reef, back-reef, and fore-reef types may be found in patchy distribution according to depth conditions *et cetera;* however, the fore-reef zones are clearly recognizable round the basinal margins of a bank-reef complex (Fig. 14).

[13] The terms "transgressive" and "regressive" are used here in the sense of movement respectively toward and away from the adjacent land area.

FIG. 14.—Pir-i-Mugrun Mountain, near Sulaimania, northeastern Iraq, showing passage of Lower-

Space does not permit the discussion of other examples, or of the many special conditions which may modify the ideal pattern of reef-complexes; these are easily visualized, but enough has been said to show that careful study of facies-zones, particularly in thin sections, may enable us to locate and reconstruct reef-complexes approximately from observation of scattered local sections, if these are adequate in number, scope, and distribution; and that much may be learned of the tectonic history of a reef by the same criteria.

It is scarcely necessary to add that a large amount of material from any province must be examined before the observer can become practiced in evaluating microscopic characters, and in assessing the significance of greater or less differences in terms of depositional environments.

It does not follow that the conditions described in the Cretaceous and Tertiary of the Middle East have exact parallels elsewhere, or in deposits of different ages. The stratigraphical setting in which the reefs occur is obviously an important factor in their development, and in that of associated sediments.

DIAGENESIS: POROSITY AND PERMEABILITY

Various opinions have been expressed concerning the alteration of reef-limestones and the writer has only to add the following observations, and their possible interpretations, relating to reef-complexes in the Middle East.

(1) Back-reef limestones (miliolid-peneroplid facies), deposited, as they are believed to be, in a zone of rapid crystallization of lime from solution, are typically dense and non-porous and are consequently very resistant to subsequent alteration by penetrating fluids.

Iiddle Cretaceous reef limestone (a), through transition beds (b), into globigerinal facies (c).

(2) Much the same is true of reef-limestone matrix which is ordinarily a dense, fine-grained, calcite cement doubtless deposited through physico-chemical and biochemical (decay) processes.

However, dead corals, rudists, *et cetera*, which escape internal cementation, retain their natural porous, cellular structure, giving access to invading fluids and rendering the rock liable to diagenetic action.

Support for this assumption has been found in some rudist reefs of northern Iraq where patchy alteration in general follows the distribution of the rudist colonies; elsewhere in the same outcrop these organisms are tightly cemented inside and outside and the whole rock is unaltered.

It is, perhaps, for this reason that many rocks full of corals and rudists are highly recrystallized. Algae, hence algal limestones, are much finer in texture, are less liable to attack, and are therefore much less altered as a rule. This accords with experience in the Middle East.

(3) Fore-reef detrital limestones are naturally porous at the time of deposition and may remain so if laid down below the zone of rapid lime-crystallization.

Such modes of origin are postulated for highly porous fore-reef limestones which show little trace either of interstitial calcite cement, or of alteration of their constituent material.

Examples occur in the Middle East.

On the other hand, a high initial porosity renders these limestones particularly liable to attack by invading fluids in a changed physico-chemical environment resulting from some later tectonic or erosional history, involving too many possible variables for discussion here.

Such changes are exemplified in fore-reef limestones which are more or less recrystallized, or which have undergone secondary cementation, which is recognizable as such because the detrital constituents are altered.

If deposited in shallow shoals within the zone of rapid lime-crystallization, fore-reef limestones may be cemented as solidly as the back-reef limestones, and their fossils and detrital elements may remain entirely unaltered. Many such limestones are known in the Middle East.

(4) Intrinsically impervious limestones undergo solution-channelling and superficial recrystallization, but not deep alteration, in the zone of percolation at a bare erosion surface.

They may, perhaps, suffer deeper alteration by exposure to soil acids, or by long immersion in the zone of permeation, at or below water-table level, according to local physico-chemical conditions.

Even in such circumstances it is the intrinsically porous, coralline and detrital fore-reef limestones which undergo the most intensive diagenesis.

(5) Many reef-limestone complexes in the Middle East reveal evidence of alternating phases of leaching and re-cementation.

Since each of these phases requires water-saturation of the limestone, or at least water with oil, such observations have an important bearing on the circumstances and date of entry of oil, if any.

Thus, in the productive limestone of Kirkuk, some cores show lime-sealed cavities containing oil which is of very different quality from that now filling the reservoir.

This observation seems to indicate an earlier phase of oil-entry, followed by water-flushing and some cementation, before the invasion of the reservoir by its present contents.

It is obvious that such cases involve so many possible variations that each local problem must be studied independently, and no useful generalizations can be attempted.

(6) Where chemical analyses have been carried out (at Kirkuk) the dolomite content of limestones in the various reef-facies-zones is extremely variable and bears no consistent relationship to porosity or alteration; some of the least porous beds are dolomitic and some of the most porous limestones have low magnesium content. It seems impossible to differentiate macroscopically crystalline dolomitic and calcitic limestones.

RELATIONSHIPS OF REEFS TO OIL GENERATION, MIGRATION, AND ACCUMULATION

Two spectacular outcropping reefs in northeastern Iraq form the basis of discussion under this heading.

One, at Pir-i-Mugrun, is Lower-Middle Cretaceous in age and is illustrated in Figure 14.

Massive, reef-like limestones (Fig. 14-a) are seen passing through an intermediate zone of interdigitation (fore-reef and fore-reef transition, Fig. 14-b) into globigerinal limestones and marls (Fig. 14-c) of basinal facies.

Microscopic examination shows that, in the transition zone (Fig. 14-b), the globigerinal beds alternate with micro-detrital fore-reef limestones containing rudist breccia, *Orbitolina* spp., *et cetera*. The massive limestone (Fig. 14-a) is too much recrystallized to reveal its original characters, but the same formation in adjacent sections is full of rudist reefs, the former existence of which at Pir-i-Mugrun is suggested by detritus in the fore-reef limestones.

The whole section is therefore interpreted as illustrating the passage from an original rudist bank-reef complex, into basinal deposits.

The latter are highly bituminous in part, the bitumen content decreasing away from the reef.

The fore-reef limestones (Fig. 14-b), where porous, are also bituminous, and so are the massive limestones (Fig. 14-a).

The other great reef formation is Upper Cretaceous in age and is typically exposed from the Bekhme Gorge to Aqra and beyond; it is also a bank deposit and consists of massive rudist reefs with associated detrital, shoal limestones in rather confused relationships and with varied foraminiferal faunas. The latter include assemblages in which there is a predominance of Miliolidae, Peneroplidae, and *Dicyclina*, or of *Orbitoides*, *Omphalocyclus*, and *Loftusia*, or of Valvulinidae (aff. *Pseudochrysalidina*), and Rotaliidae.

However, at the bank margin the complex passes through long tongues of fore-reef shoal limestones with *Loftusia*, *Omphalocyclus*, *Orbitoides*, *et cetera*, and with some reef patches, into basinal deposits which in this case are partly in a globigerinal marl, and partly in an arenaceous, facies.

Intermediate stages in the passage from massive reef-limestones into bedded fore-reef limestones with orbitoids can be observed at Aqra.

The reef and fore-reef limestones are saturated with bitumen wherever the rock has primary or secondary porosity, and the adjoining basinal deposits are also bituminous in part.

It requires little imagination to suppose that oil was generated in the richly organic basinal depths in front of these reefs, where the latter doubtless gave some protection from aerated currents, thereby helping to preserve an euxinic bottom-environment.

As oil was pressed out by compaction, it found a ready outlet along the bedding through porous, fore-reef carrier beds, closely interfingering with the source rocks, which led it up to the reef-reservoir. There much of it no doubt escaped, but some would be trapped in irregular pockets and porous zones within the dense matrix, or would be covered by cemented, fore-reef limestones deposited in the higher, shallower shoals in front of the reef. A primary stratigraphic accumulation of oil would be formed.

Later, when folding and faulting occurred, fractures would permit this oil to spread through the reef-complex, and to escape, unless the whole had been covered by later, impervious deposits. In the latter case, secondary concentration would follow in any accessible, structural trap, and an oil field would result.

At Kirkuk, the fore-reef shoal limestones make up the thickest and most porous parts of the reservoir; the reef and back-reef limestones yield only fissure production.

It is obvious that many variations of this story can be visualized. It is quite possible that some oil was actually formed in reef and fore-reef limestones. Interesting examples have been observed of well preserved rudists filled with bitumen but apparently embedded entirely in impervious, unaltered limestone matrix which, however, smells of oil on fracturing.

But supporters of theories postulating slow, multi-stage conversion of organic matter into petroleum, may prefer to regard the movement of oil into a reef-reservoir as a long continued process under tectonic or other impulses.

The writer believes that critical study of the reefs described might yield valuable data bearing on these problems, and it is hoped that the necessary investigations may be undertaken at some later date.

Most of the exposed reef-complexes in the Middle East show no signs of oil. In some examples, flushing out on subsequent exposure is established, but other cases suggest that some or all of the following conditions may account for the absence of oil. 1. The reefs and adjoining basins were too shallow or too current-swept to favor anaerobic bottom conditions for generation of oil. 2. The fore-reef limestones were deposited in shoals at shallow depths; hence they were cemented by primary lime-precipitation. 3. Cover and stratigraphic trap conditions did not exist before folding so that any oil formed was lost.

Reprinted from
BULLETIN OF THE AMERICAN ASSOCIATION OF PETROLEUM GEOLOGISTS
VOL. 34, NO. 2 (FEBRUARY, 1950), PP. 263-294, 18 FIGS., 2 PLATES

THEORY OF TRANSGRESSIVE AND REGRESSIVE REEF (BIOHERM) DEVELOPMENT AND ORIGIN OF OIL[1]

THEODORE A. LINK[2]
Calgary, Alberta, and Toronto, Ontario, Canada

ABSTRACT

PART I

The established geological principle of transgression and regression of epi-continental seas, the resultant sediments, and their fauna and flora, is applied to coral-reef or bioherm-forming organisms. ("Coral reef" and "bioherm" are regarded as synonymous.) Bioherms which develop during a transgression are differentiated from those of a retreating sea by the associated sediments. A transgressive bioherm is surrounded and overlain by marine clastics deposited during submergence, while the regressive type of bioherm is associated with evaporites and/or other types of sediments deposited during withdrawal of the sea.

PART II

It is suggested that hydrocarbons found within coral-reef or bioherm reservoirs are indigenous, because of the obvious concentration and accumulation of organisms in them. The porosity and permeability of coral-reef or bioherm reservoirs are attributed not only to the hollow corallites *et cetera*, but also to the helter-skelter accumulation of them so that, in many instances, such porosity is greater, more effective, and more continuous. Partial or entire obliteration of porosity is, in part, due to infiltration of evaporites associated with the regressive type of bioherm.

DEFINITION OF TERMS "BIOHERM" AND "BIOSTROME"

A "bioherm" is an accumulation of the secreted hard parts of sessile types of primitive invertebrates and plants such as corals, bryozoans, crinoids, sponges, algae, *et cetera*, which accumulation is greater than the surrounding contemporaneously deposited sediments, thus giving rise to mounds, ridges, or reefs. Therefore, "coral-reefs" and "bioherms" are, in a sense, one and the same. According to what organism predominated to cause it, such an accumulation is termed a "coral-reef" or "coral bioherm"; a "bryozoan reef" or "bryozoan bioherm"; an "algal reef" or "algal bioherm," *et cetera*. When in doubt one should refer to them as "bioherms" or "reefs" without commitment as to composition.

A "biostrome" is an accumulation of the same or similar sort of material as a "bioherm" or "reef," but the accumulations are in layers or strata which are no greater in vertical dimensions than the contemporaneously deposited surrounding sediments. Obviously there is a gradation zone between biostromes and bioherms for, no matter how thin or how thick the accumulation may be, it pinches or lenses out somewhere, and thus any biostrome must be regarded as a lens even though it may be only 2 feet thick and 100 miles or more in areal extent. Also, it may have irregularities or humps on top of it, of varying sizes, which could be termed "bioherms." One could, in a sense, look upon these primitive bioherm- or biostrome-forming organisms as catalysts in the sea which cause excessive precipitation or concentration of carbonates from the sea-water wherever the environment is favorable.

PART I

GENERAL DISCUSSION OF BIOHERM (CORAL-REEF) AND BIOSTROME DEVELOPMENT

Studies by naturalists,[3] biologists, and geologists of recent coral reefs,[4] or bioherms,[5] indicate that the following are some of the conditions which govern

[1] Read before the Association at St. Louis, March 16, 1949, as part of the symposium on reefs. Manuscript received, August 31, 1949.

[2] Consulting geologist, 810A First Street West, Calgary, Alberta, and 31st Floor, 25 King Street West, Toronto, Ontario, Canada.

their growth and existence: (1) depth of water—25 fathoms, (2) good circulation, (3) salinity, 27–38 parts per thousand, (4) continuous submergence excepting at extremely low tide.[6] However, there is no doubt that the most suitable water depth for all genera or species of any bioherm is not the same, and this also applies to the other factors enumerated. It is therefore logical to conclude that a bioherm composed of one ecological assemblage would survive where another, with a different assemblage, would not; furthermore, a bioherm with a definite assemblage could continue to survive even though a part of its assemblage might die out. This sort of reasoning could be carried still further by imagining that a change of conditions might eliminate one part of the assemblage for a period, but it would return when conditions revert to the original; or still further, one part of the assemblage could become dwarfed or depauperized and return to normal after original conditions return. One could go on with this reasoning and finally arrive at the conclusion that the conditions governing the growth and extinction of a bioherm are not as fixed as one is led to believe on the basis of the study of living reefs or bioherms. The study of ancient bioherms verifies this conclusion. Those of Ordovician time did not have the same ecological assemblage as those of Permian time, nor do either correspond with those of the present. Furthermore, the overlying, contiguous, or associated sediments of bioherms vary within a geologic age, as their assemblage of organisms varies.

Interesting speculation along these lines would be an explanation of the Devonian bioherms far north, inside the Arctic Circle in northwestern Canada, if abundant sunlight and warm waters were a necessity. If it is assumed that a warm climate did exist during Upper Devonian time in this latitude, it could be explained by tropical currents from the south, but the abundant year-round sunlight would call for a tilting of the axis of the earth to make possible the intensity of sunlight which recent reef study postulates. If there had been a tilting of the global axis it would need fit the Arctic reefs in north Canada and the Devonian reefs as far south as New York and southwestern Texas. The global axis might have been almost at right angle to what it is to-day in order to have the same amount of sunlight in southwestern Texas and Arctic Canada at the same time. It is because of this kind of reasoning that one is tempted to challenge the conclusion of Darwin and many present-day geologists and naturalists that coral reefs grew only during submergence and under conditions as observed to-day. In many cases, the data from the geological column contradict such conclusions. The most obvious evidence that reefs were developed during withdrawal of an

---

[3] Charles Darwin, *Structure and Distribution of Coral Reefs*. 344 pp. D. Appleton and Company, New York (1897).

[4] The terms "coral reef" and "bioherm" are regarded as synonymous and applied as suggested by W. B. Wilson, chairman of St. Louis program committee.

[5] In this contribution the terms "bioherm" and "biostrome" are employed as defined by E. R. Cumings, "Reefs or Bioherms," *Bull. Geol. Soc. America*, Vol. 43 (1932), pp. 331–52.

[6] T. W. Vaughan, "Geology of Modern Marine Organisms with Reference to Paleogeography," *Bull. Geol. Soc. America*, Vol. 51 (1940), pp. 433–68.

epi-continental sea are the impressive, mountain-size outcrops of the El Capitan reef in West Texas and New Mexico, and the most amazing thing about practically all modern text-books on sedimentation, stratigraphy, and general geology is the conspicuousness of this classic area by its entire lack of mention.[7]

DEVELOPMENT OF A BIOSTROME[8]

As already stated, lime-secreting organisms such as coral, bryozoans, and algae are known to require special conditions or environments in which they thrive, and these conditions are not necessarily fixed but lie within certain ranges, so that there is some leeway if some of the conditions are not fully obtained. Furthermore, if the environment is not altered too suddenly some of the organisms are capable of adjusting themselves within reasonable limits to the gradual change. However, if the conditions or the environment change beyond the capacity of the organism or organism assemblage to become adjusted, they will either become extinct or migrate if possible with the advancing or retreating sea.

Let us assume an epi-continental sea within which exists a favorable environment conducive to the growth of lime-secreting organisms such as corals, bryozoans, crinoids, and algae, and let us assume further that this environment is fairly limited in vertical range (depth below the sea surface) as indicated in Figure 1. Under such condition the type of organism in question which invaded the area, and indicated as X in Figure 1 (Stage I), should eventually spread over the sea-bottom as shown in Stage II, under the assumption that the sea-level remains stationary within the limits required of this organism X, or assemblage X. The accumulation of the secreted hard parts of this specific assemblage would be limited to the vertical range indicated in the diagram, and eventually there would result a stratum or layer of these coralline hard parts as indicated in Stage III. Once the accumulation was so thick that the depth below the postulated stationary sea-level was not great enough, this particular group of organism would cease growing and accumulating vertically in this area and would either become extinct or migrate into a favorable area, and thus give rise to a blanket deposit of coralline limestone termed a "biostrome." Eventually, the biostrome would become covered by later sediments, but if a recurrence of this favorable environment should take place in the area, this could give rise to successive layers, or biostromes, of coralline rock which would be described as rhythmic or cyclic deposits (Stage IV, Fig. 1).

Excellent examples of such biostromes may be observed in the Ordovician limestone quarries near Winnipeg, Manitoba, where the Tyndal or mottled limestone is quarried, and from which bedded rocks a great many buildings such as post offices, city halls, and hotels were constructed or ornamented in western Canada. Accumulations of oil and/or gas in such reservoir rock in large concen-

[7] P. B. King, "Geology of the Southern Guadalupe Mountains," *U. S. Geol. Survey Prof. Paper 215* (1948).

[8] E. R. Cumings, "Reefs or Bioherms," *Bull. Geol. Soc. America*, Vol. 43 (1932), pp. 331–52.

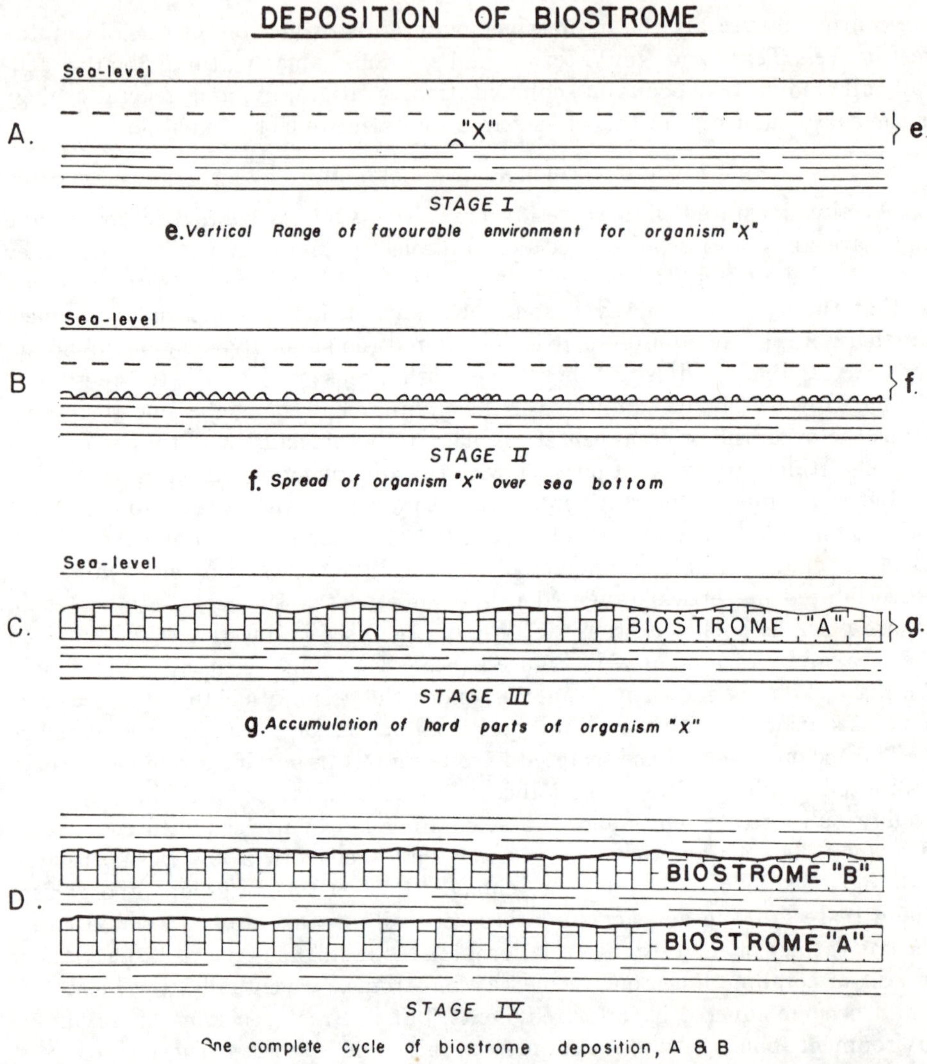

Fig. 1.—Ideal diagrams illustrating growth of biostrome A on sea-bottom within favorable vertical range of organisms or assemblage of organisms, and recurrence of same conditions as biostrome B.

trated amounts would require structural traps similar to a blanket sand deposit, or lens of considerable areal extent.

Biostromes may also be developed because of entirely different reasons, such as inability of certain types of lime-secreting organisms to grow any considerable amount vertically because of their fragile nature and tendency to thrive or grow in almost horizontal positions, for example, some frail bryozoans and corals.

Figure 2 shows a development of a biostrome like that in the previous figure.

but a differential downwarping or tilting is postulated as having taken place without change of sea-level toward the end of Stage III. Under the assumption that the sea-level remained stationary, the favorable environment for the biostrome-forming organism continued as shown in Stage IV at the center, and extended toward the left, causing growth of the coralline material in its suitable environment, but not in the deeper waters at the right. Thus there may have developed a bioherm on top of the biostrome, which could act as a stratigraphic trap for accumulation of oil and/or gas. Continued downwarping would obviously

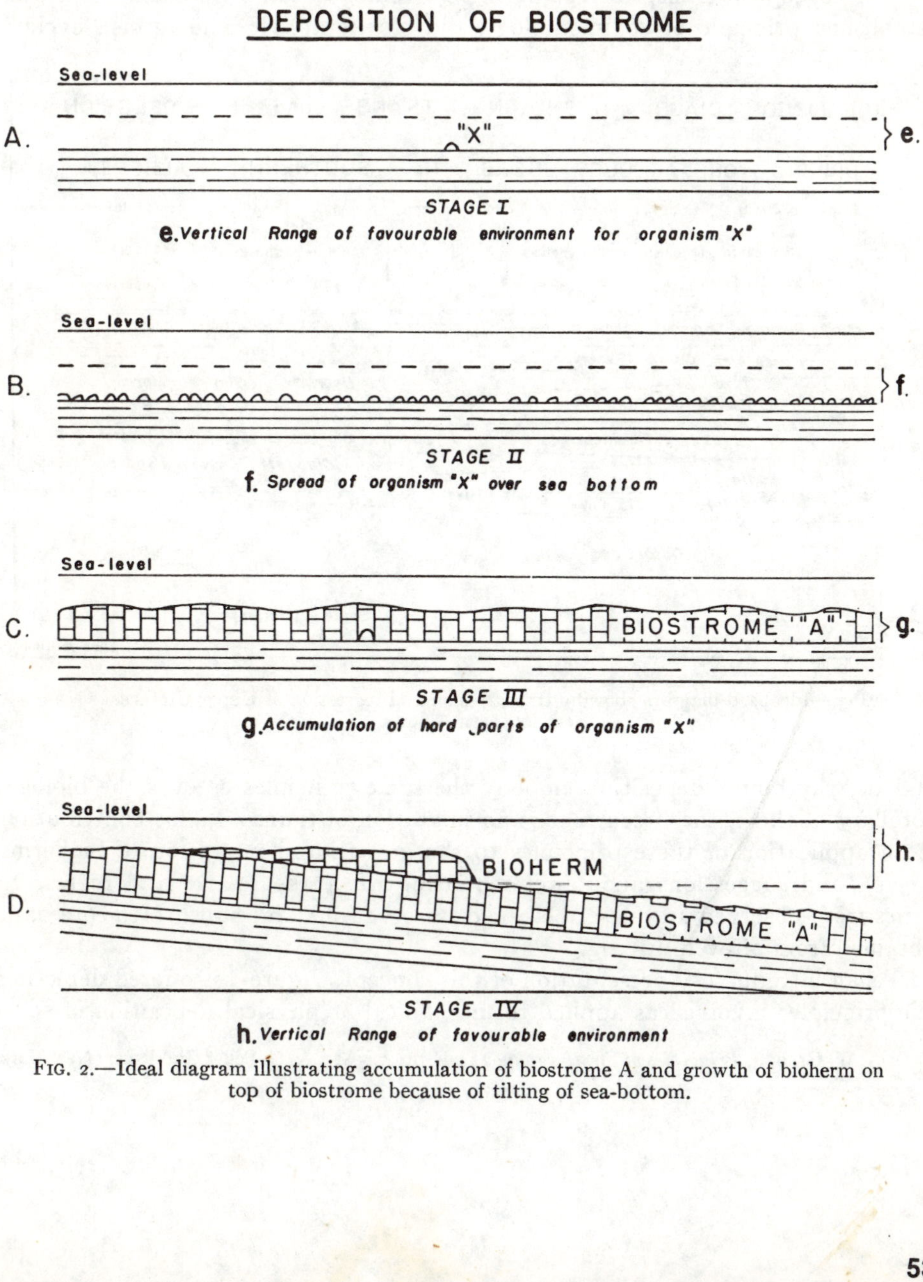

FIG. 2.—Ideal diagram illustrating accumulation of biostrome A and growth of bioherm on top of biostrome because of tilting of sea-bottom.

cause further vertical and shoreward growth of the bioherm, and this leads to consideration of other conditions which may cause the concentration of the remains of lime-secreting organisms in mounds or reefs and thus give rise to the growth and migration of bioherms.

TRANSGRESSIVE AND REGRESSIVE BIOHERM DEVELOPMENT (SUGGESTED HYPOTHESIS)

GENERAL DISCUSSION

The concept of transgressive and regressive deposition of sediments is a well established principle of sedimentology and stratigraphy.[9] The terms "cyclic"

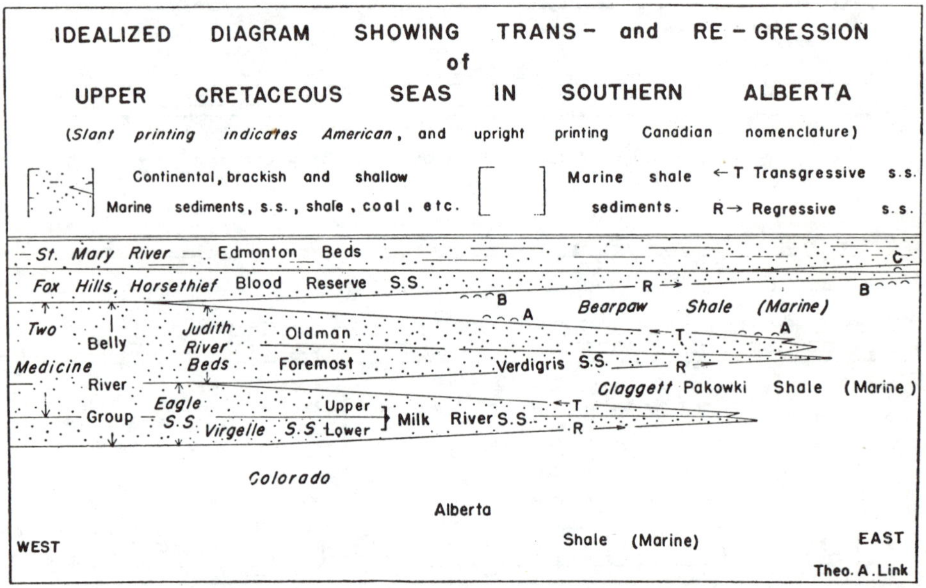

FIG. 3.—Idealized diagram showing transgression and regression of Upper Cretaceous seas in southern Alberta.

and/or "rhythmic" deposition embody the same principles as does the biologic corollary of these, the "recurrence and migration of faunas." This contribution is an application of these principles to the growth and migration of bioherms (reefs), and suggestions about differentiating the transgressive and regressive types from observed and described field observations and subsurface core data obtained from oil-well drilling.

As an introductory presentation of this concept, Figure 3 is offered depicting the principles involved as applied to mechanical or physical deposition of sedi-

[9] A. W. Grabau, *Principles of Stratigraphy*, 2d ed., pp. 725–45. A. G. Seller, Publisher, New York (1924).

ments as well as to the biological aspect. The diagram illustrates the recurrent marine invasions and withdrawals in southern Alberta during late Upper Cretaceous time, giving rise to brackish and continental sand wedges interfingering with marine shale. As a result there are transgressive as well as regressive sand and shale deposits with their appropriate faunas. Observation of these conditions was made by the writer[10] while conducting detailed field work in that area 20 years ago, and the recurrence of marine oyster deposits above and below the sand wedges or fingers illustrated the principle of migration and recurrence of a fauna so clearly that publication of this has been in mind ever since. At the base of the Bearpaw shale at A is a layer of marine oysters above a thin section of marine shale overlying the sandstones, coal, and brackish shale of the Belly River series. The same or similar oysters also occurred in the near-shore marine shale directly below the Fox Hills sandstone as indicated at B in the diagram. The Bearpaw shale between these two occurrences of oysters contains deeper marine sediments and a corresponding fauna containing baculites (*Placenticeras meeki*), thus indicating that the *Ostrea* fauna did not migrate up the geological column vertically, but evidently followed the transgression of the Bearpaw sea westward, and then followed it out eastward during regression. Since the oysters are also found above the Fox Hills[11] sandstone at C, it is evident that a westward transgression of the sea again occurred in the area after deposition of the lower Fox Hills member.

The correct interpretation of the occurrence of the several *Ostrea* beds is apparent on this diagram, which covers miles of area and is the result of much field observation. However, let it be assumed that the field geologist was able to observe only a very limited outcrop of the *Ostrea* beds, with no data about sediments above or below them. He could easily miscorrelate bed A with bed B on fossil and lithological evidence as time and stratigraphic equivalents. If enough vertical section were exposed in both instances he would observe that bed A overlies sands, coal, and other brackish-water or near-shore sediments, and is overlain by marine shale. This would be enough evidence to suggest a transgression of the sea or a rising strand line. In the case of bed B he would find the reverse, that is, marine shale below the oyster bed and near-shore sandstones above it, thus suggesting a lowering of the strand line or a regressive cycle as already stated. This contribution is merely an application of these fundamental geological principles to reef or bioherm growth which, as an experienced geologist should know, rarely follows idealized patterns.

In general, the sessile types of organisms which give rise to the formation of bioherms propagate by budding[12] or stolonation as well as by sexual reproduction. Therefore, when a sea invades an area most of the bioherm-forming organisms must advance or shift with the waters of the sea by means of mobile, fertilized

[10] Theodore A. Link and A. J. Childerhose, "Bearpaw Shale and Contiguous Formations in Lethbridge Area, Alberta," *Bull. Amer. Assoc. Petrol. Geol.*, Vol. 15, No. 10 (October, 1931), pp. 1227–42.

[11] J. O. G. Sanderson, "Fox Hills Formation in Southern Alberta," *ibid.*, pp. 1251–63.

[12] T. W. Vaughan, *op. cit.*

ova, planulae, ciliated larvae, *et cetera*. Furthermore, the bioherm-forming organisms are not nearly so mobile and adaptable as the planktonic type, so that in the case of very rapid advance or retreat of an epi-continental sea they may not survive to accumulate in mounds. However, each ciliated larva which takes anchor is a potential incipient bioherm if, by chance, it takes root in a favorable environment. Obviously, there must be considerable latitude in the environment for each type of bioherm-forming organisms, be it temperature, depth below the surface of the water, circulating or stagnant water, or salinity. If this were not so, the development or growth of bioherms would not be so com-

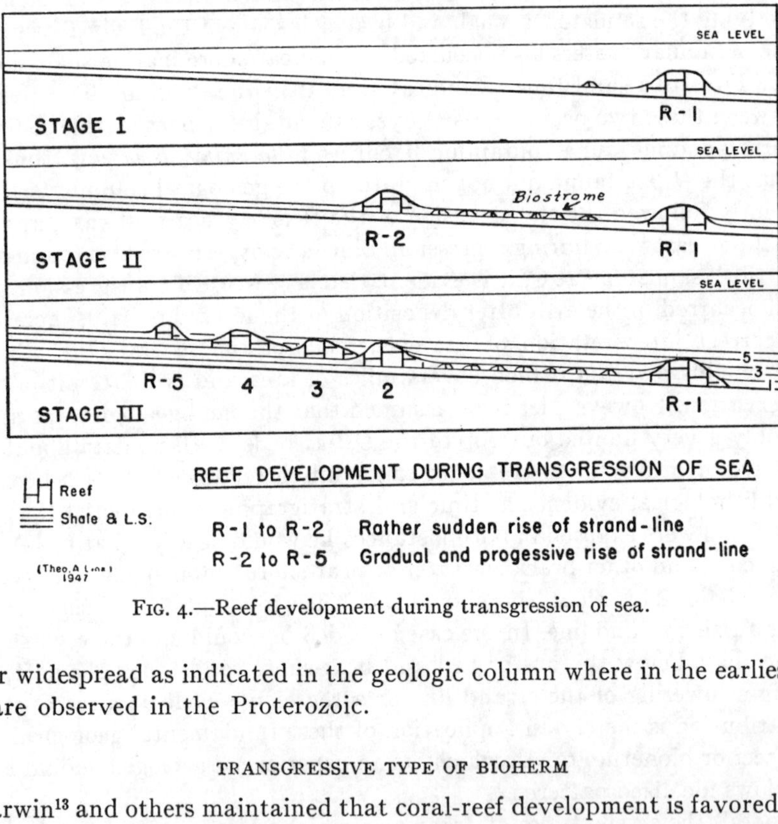

FIG. 4.—Reef development during transgression of sea.

mon or widespread as indicated in the geologic column where in the earliest algal reefs are observed in the Proterozoic.

TRANSGRESSIVE TYPE OF BIOHERM

Darwin[13] and others maintained that coral-reef development is favored during the rise of the strand line, and he went to great lengths to prove his point with reference to atolls and fringing reefs. However, the records of geologic history indicate that reefs or bioherms developed during both, advance and retreat, of strand lines.

Figure 4, Stage I, is an idealized hypothetical case of a bioherm (R-1) growing upward because of a gradual rise in sea-level, such as postulated by Darwin.

[13] Charles Darwin, *op. cit.*

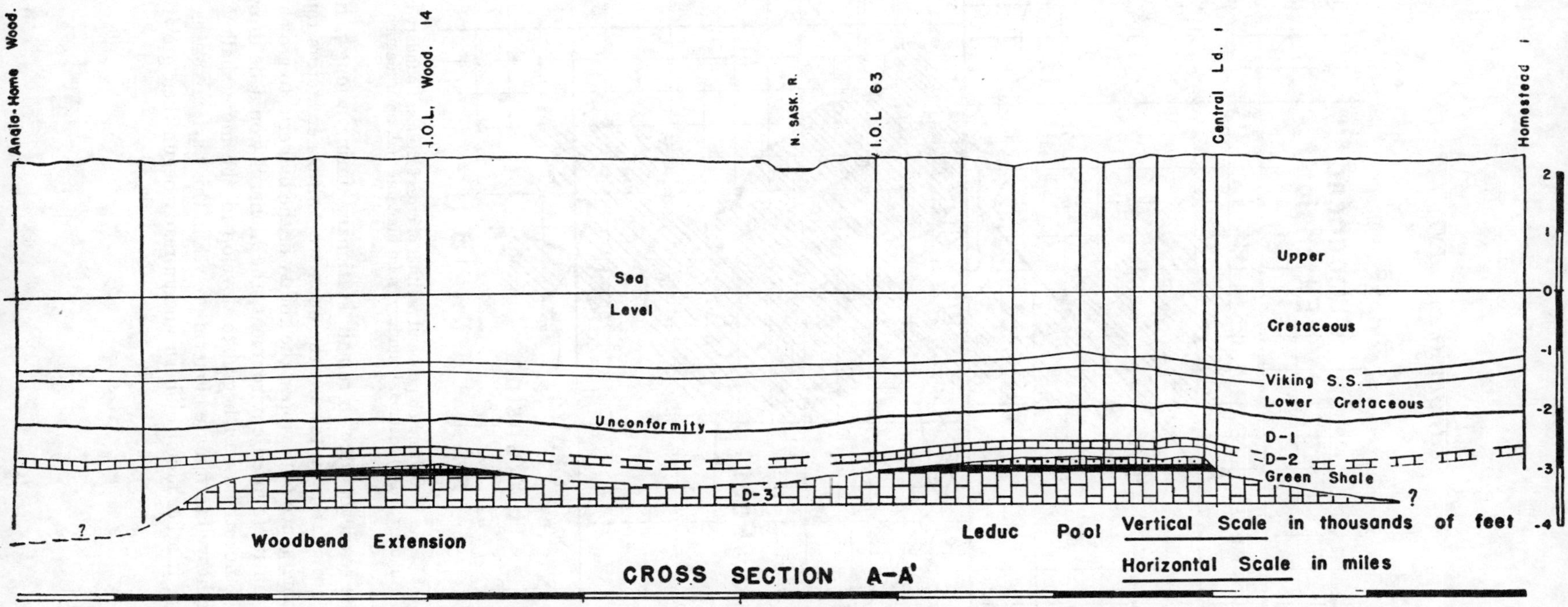

FIG. 5.—Cross section through north tip of Leduc oil pool and Woodbend extension, Alberta. D_3 gas-cap shown by dots, oil column solid black, water below it. (For position cross section, see Figure 6.)

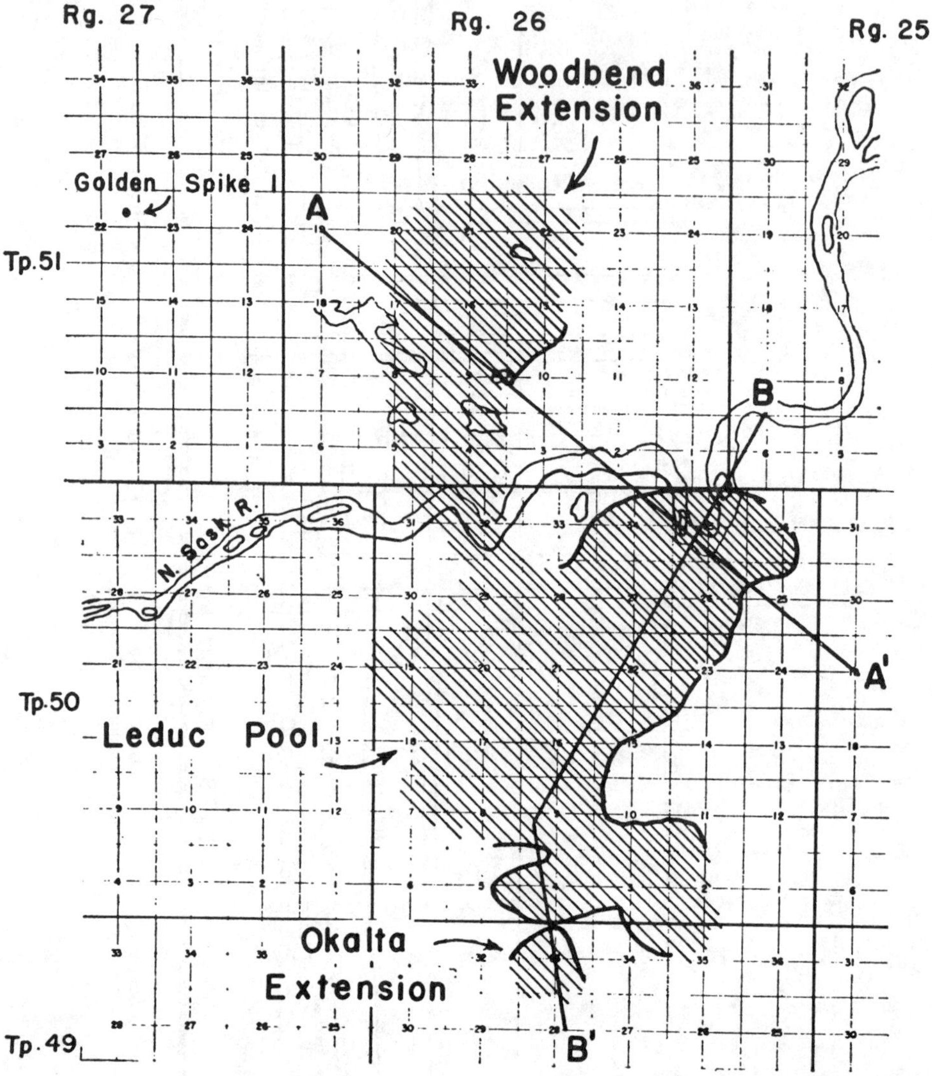

FIG. 6.—Index map of Leduc-Woodbend oil pool and position of cross sections. Boundary of oil area in D3 zone (August 1, 1949): defined boundary, solid line; inferred, open diagonals.

Obviously, if the sea-level were to remain stationary during a long period, the bioherm would grow laterally toward the open sea—all other factors remaining the same. If submergence outstripped the ability of the bioherm to grow upward, and thus maintain its favored environment, the bioherm would be drowned or silted up, and, if no others had begun to develop in the more shallow waters near-by, this might well terminate bioherm growth in this particular area for this particular period. If other minor bioherms had already begun their growth in the more favored shallow waters, some of these might eventually grow into large bioherms, or a biostrome could result as shown in Stage II on the diagram between R-1 and R-2. However, it is assumed that another bioherm did develop shoreward. In Stage II such a bioherm is indicated as R-2, which began its growth during the last stages of R-1, the latter now being dead and being covered with shales and/or limestones. Continued submergence would also eventually end the growth of bioherm R-2, and the successive bioherms indicated as R-3, R-4, and R-5 would give rise to a compound affair progressing shoreward, as shown on the illustration. With continued submergence all these bioherms would eventually be covered by shale and/or limestone, and thus be sealed completely, so that any accumulation of oil, gas, and water would be in a completely isolated reservoir.

The lower or D-3 producing zone at Leduc, Alberta[14] (Figs. 5, 6, and 7) is regarded as a reef or bioherm of this type, which grew vertically 700 feet or more, due to a progressive and gradual submergence of the area, but which was finally drowned or silted up. After complete delineation by drilling, this bioherm might be recognized as an example of a fossil atoll, whose shape could be semicircular, crescentric, elongate, *et cetera.* The history of this reef fits the Darwinian submergence theory very well. There are no evaporites within, overlying, or surrounding this bioherm.

The Norman Wells oil pool[15] in the Northwest Territories, Canada, is also a Devonian reef reservoir of approximately the same age as the Leduc D-3 zone, but like the newly discovered Redwater pool, it differs from the Leduc D-3 bioherm in that it is composed almost entirely of limestone rather than dolomite. Figure 8 is a longitudinal section of the Norman Wells reef which rests on Fort Creek shale, and is surrounded and also overlain by the same shales. It is classified as a transgressive type, having no evidence of difference in facies between the back-reef and fore-reef sediments. Anhydrite and other evaporites are lacking within the Norman Wells reef, and are absent throughout the Mackenzie River basin in the surrounding and overlying shale. Figure 9 is a transverse section of the Norman Wells reef, and its similarity to the D-3 zone of the Leduc area is indeed striking. The total maximum thickness of the Norman Wells reef is 450-500 feet, which is also comparable with the maximum thickness of 700 feet of the D-3 zone at Leduc, Alberta.

REGRESSIVE TYPE OF BIOHERM

A manner in which regressive reef growth or basinward migration could take place is illustrated in Figure 10. The first bioherm is indicated in Stage I as R-1. The back-reef or shoreward facies, indicated on the left as evaporites and clastics

[14] D. B. Layer, "Leduc Oil Field, a Devonian Coral-Reef Discovery," *Bull. Amer. Assoc. Petrol. Geol.*, Vol. 33, No. 4 (April, 1949), pp. 572–602.
Theodore A. Link, "Leduc Oil Field," *Bull. Geol. Soc. America*, Vol. 60 (1949), pp. 381–402.

[15] J. S. Stewart, "Norman Wells Oil Field, N.W.T., Canada," *Structure of Typical American Oil Fields*, Vol. III, Amer. Assoc. Petrol. Geol. (1948), pp. 86–109.

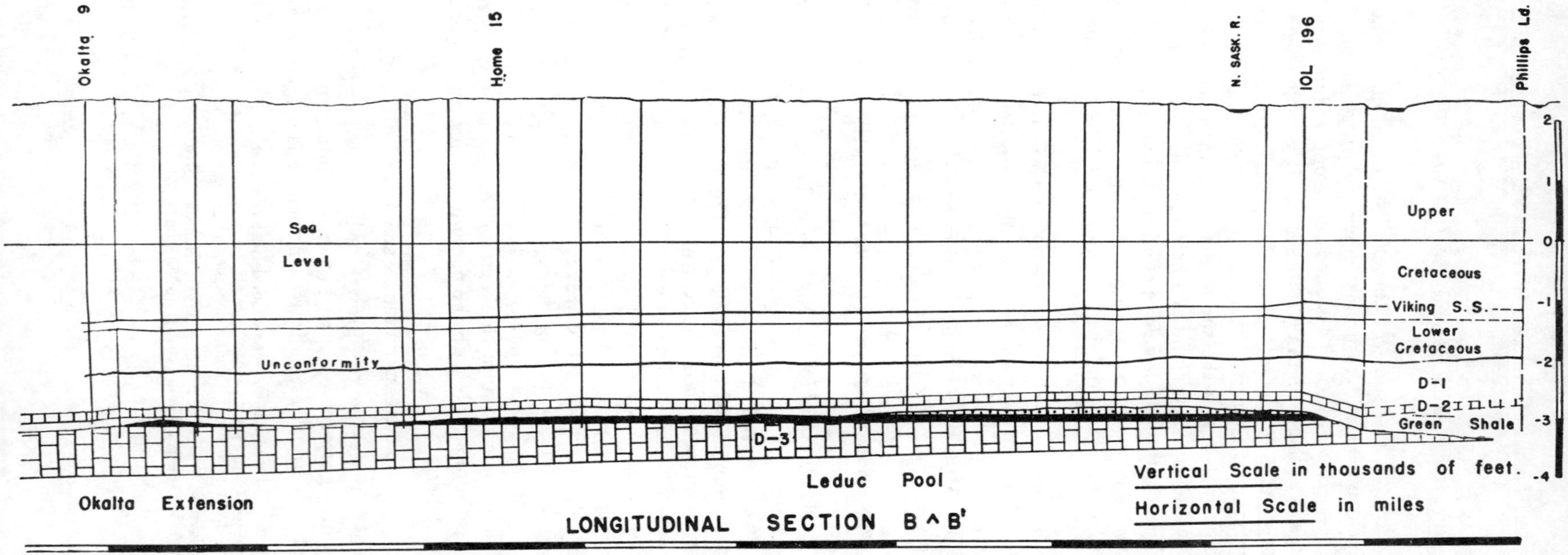

FIG. 7.—Longitudinal section through Leduc oil pool and Okalta extension, Alberta. D3 gas cap shown by dots, oil column solid black, water below it. (See Figure 6 for position of section.)

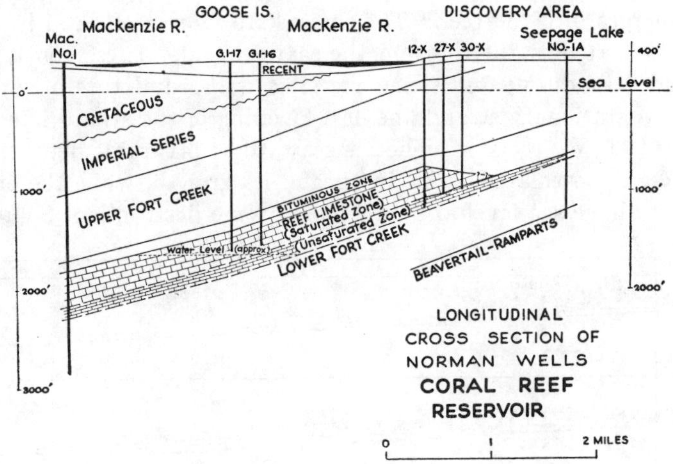

FIG. 8.—Longitudinal cross section of Norman Wells coral-reef reservoir.

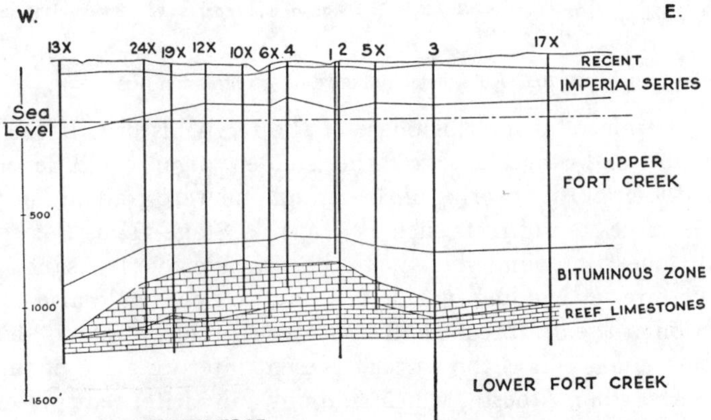

FIG. 9.—Transverse cross section of Norman Wells coral-reef reservoir.

consisting of limestones, dolomites, sandstones, red shales, anhydrites, gypsum, *et cetera*, merges into the reef.[16] The immediate "fore-reef" and basin or pontic facies are indicated at the right on the seaward side. These consist of gray to black, bedded limestone and shales merging, with admixture of sands, into the reef facies. A stationary strand line or a lowering of the strand line would cause development of new reefs or bioherms seaward at favorable depths. In Stage II bioherm R-2 is postulated as having begun its growth, while the emergent reef R-1 is at or above sea-level and undergoing some destruction. Solution breccia-

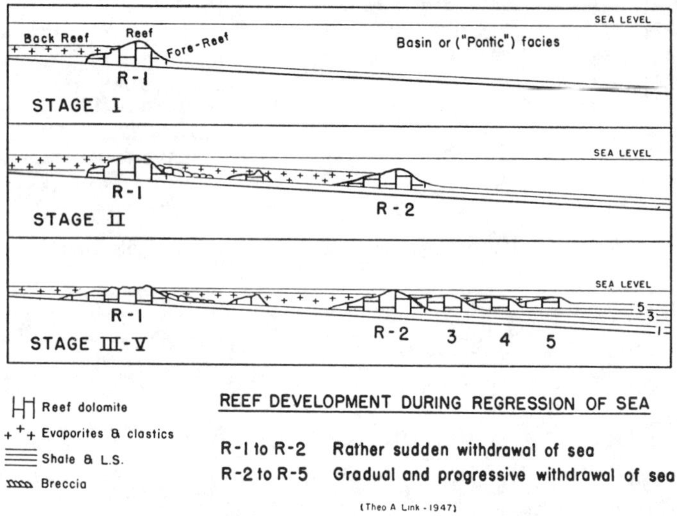

FIG. 10.—Reef development during regression of sea.

tion could take place during this phase of the reef's history. If the withdrawal between Stage I and Stage II were rather sudden, a gap could develop between reefs R-1 and R-2; but if lowering of the strand line were gradual but progressive, a continuous succession of reefs such as shown in Stage III as R-2 to R-5 might develop into one compound reef, just as regressive overlap sands cover wide areas, or like fore-set bedding in a delta. Wave action would also tend to cause the destruction of the dead reefs, and thus give rise to continuous sheets of brecciated reef material. This is the writer's present interpretation of the D-2[17] producing zone at Leduc, Alberta, which contains anhydrite, red splotches of shale and other material closely related to the back-reef and overlying facies. The D-2 zone could also be regarded as a regressive biostrome, because of its fairly uniform thickness, and great areal extent.

[16] E. Russell Lloyd, "Capitan Limestone and Associated Formations," *Bull. Amer. Assoc. Petrol. Geol.*, Vol. 13 (1929), pp. 645–58.
P. B. King, "Permian of West Texas and Southeastern New Mexico," *ibid.*, Vol. 26 (1939), pp. 535–763.

[17] Theo. A. Link, *op. cit.*

VARIATION OF SEA-BOTTOM SLOPE

Transgression.—The hypothetical development of transgressive and regressive bioherms as described thus far assumes a uniform slope of the sea-bottom, and it was suggested that concentration of reef development would, under such conditions, result only if the strand line halted from time to time sufficiently for the development of sizeable bioherms. There are many other factors which might give rise to a concentration of reef-forming organisms; it is not attempted at the present time to cover all such possibilities. However, one factor, the topographic configuration of the area being invaded by an advancing sea, might be considered. Here also, only the simplest variations are considered. An advancing

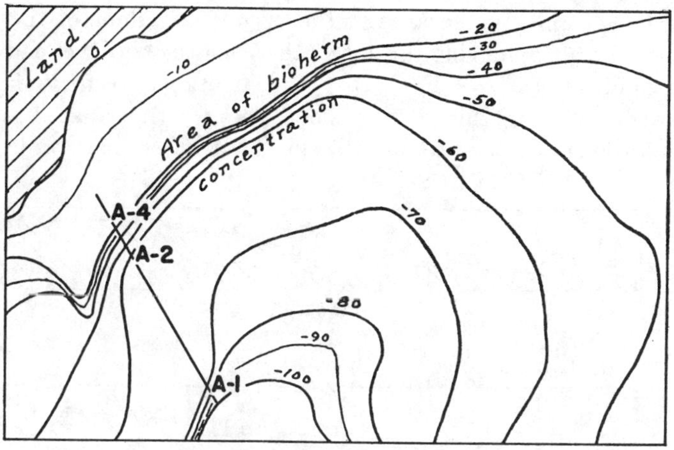

FIG. 11.—Idealized map showing configuration of basin over which sea advanced. A-2 to A-4: steeper slope.

sea, containing potential bioherm-forming organisms would, if the advance of the sea were gradual, with limited pulsations, and the floor had an even slope (all other things being equal), give rise to an even layer of reef-forming material such as termed the "biostrome" by Cumings.[18] If, however, the sea-floor which is being invaded were not of a constant gradient, but had steeper slopes here and there, or jutting promontories, as indicated in Figure 11, a concentration of the bioherm-forming organisms should take place at such changes of slope regardless of a uniform, non-pulsating advance of the sea. This is illustrated by cross sections in Figure 12, on which are shown two stages of an advancing sea in which a transgressive development of bioherms is postulated. It is assumed that during the last stages of growth of bioherm A-1 a progressive rise of the strand line took place, that the bioherm A-1 could not maintain its growth, and died. From Stage I to Stage II there would be little possibility of large-scale bioherm development

[18] E. R. Cumings, *op. cit.*

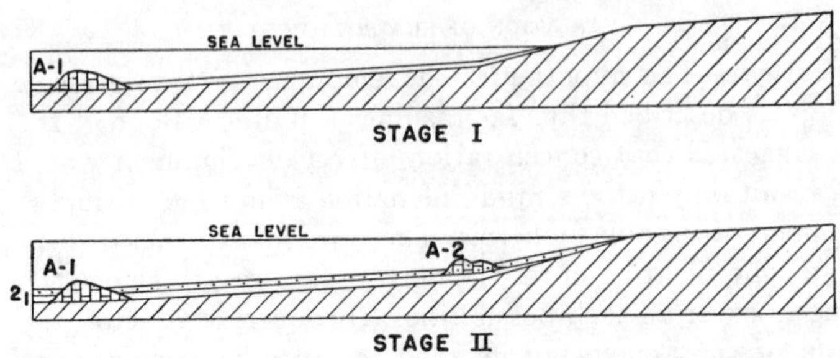

FIG. 12.—Ideal diagram showing concentration of transgressive type of bioherms on steeper slope during advance of sea. Stages I to II.

between reefs A-1 and A-2 because of the very gentle slope of the sea-floor. A biostrome could develop during this interval. When the sea-level had reached the steeper slope indicated at A-2 (Fig. 13), a concentration of reef-forming organisms should take place along this steeper slope because the ciliated larvae would require less distance to travel to find the suitable environment, and thus more of

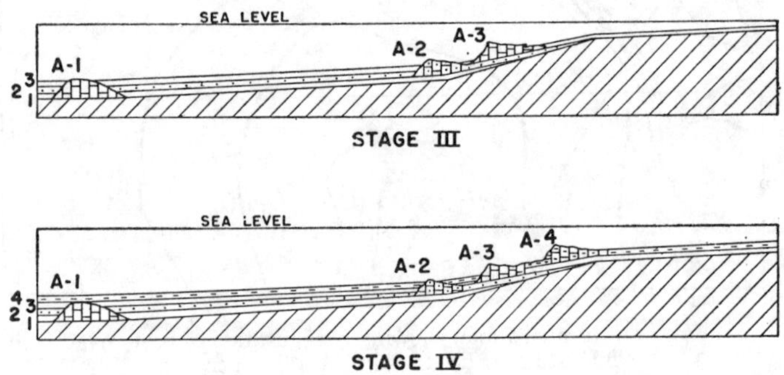

FIG. 13.—Ideal diagram showing concentration of transgressive type of bioherms on steeper slope during advance of sea. Stages III to IV.

them would survive and begin budding sooner. Also, during an equal interval of time and an equal rise in sea-level, more opportunity for growth of the bioherm by stolonation or budding is afforded, because the lateral growth and accumulation of bioherm material would be confined. This applies from Stage II to III, and to Stage IV. Thus, all other factors being equal, there should be a concentration of bioherm accumulation on the steeper slope.

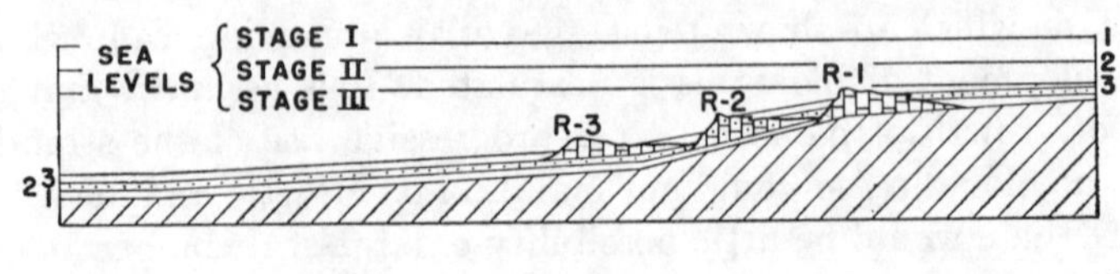

FIG. 14.—Ideal section showing development of regressive bioherms (reefs) concentrated on steeper slope.

Regression.—The same reasoning employed above also applies in the case of bioherms developed during the regression of a sea, as indicated on Figure 14. In this illustration the various stages of regression have not been shown individually, but the final theoretical picture of the three stages is illustrated as having been concentrated along the steeper slope, with the back-reef or shoreward facies at the right. Due to lowering of the sea-level, some destruction of the earlier reefs by wave action would probably have taken place, thus giving rise to reef breccias, as well as bevelling-off of the tops of the reefs.

Transgression and regression.—On the basis of the foregoing discussions and illustrations, Figure 15 has been prepared as an ideal diagram to depict the devel-

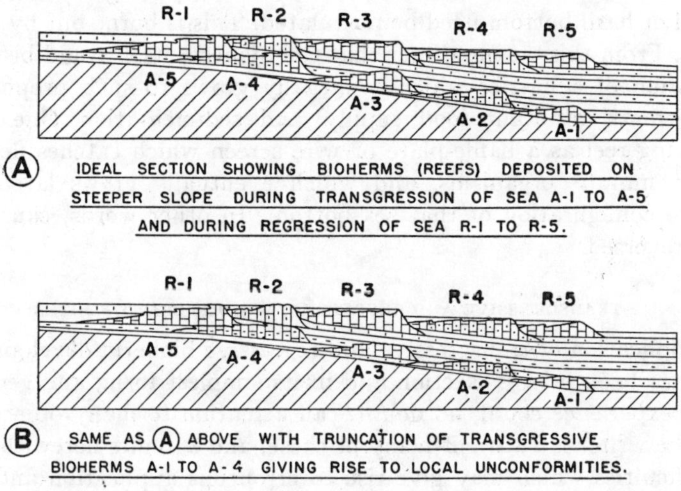

FIG. 15.—Ideal section showing bioherms (reefs) deposited on steeper slope during transgression of sea.

opment of a series of transgressive (A-1 to A-5) bioherms, showing the maximum advance of an invading sea, and then the development of a series of regressive (R-1 to R-5) bioherms resulting from the retreat of the sea. In the illustrations the bioherm indicated as A-5 and R-1 is regarded as one and the same, and it represents the one developed during the last stages of the maximum advance of the sea and the first stage of the retreating sea, and could be regarded as a barrier reef if of sufficient linear extent. Behind it, or shoreward, the back-reef facies were being deposited contemporaneously with the reef development as well as the fore-reef and pontic facies.

During the advance of the sea, the transgressive bioherms A-1 to A-4 might have undergone some submarine denudation or destruction, dependent on the wave-base and rate of submergence, thus causing the tops of the bioherms to be bevelled off before deposition of the succeeding clastic sediments. This would give rise to what might be mistaken for large-scale unconformities as indicated in the lower part B of the illustration. Such false or apparent unconformities might show a progressive upward shift in the stratigraphic section. Field observations of such a local condition could give rise to an erroneous conclusion that a large-scale major unconformity is under observation. Possibly some of the West Texas and New Mexico Permian correlation disagreements between stratigraphers and paleontologists might be due to conditions analogous with those herein described, and an application of these simple principles could conceivably eliminate some of the present disagreements of correlation.

Umbgrove and Kuenen[19] demonstrated that bioherms can begin their growth on a soft, muddy, stable bottom, if the other factors or conditions are satisfied. Therefore, it is conceivable that no platform or steepening of sea-bottom slope, nor a solid or hard bottom need be postulated. This is borne out by the Norman Wells reef. From this we may infer that the bioherm itself could be the cause of the formation of a barrier which eventually grows to such proportions as to change currents, resultant food supply, and sedimentation. One might look upon a living reef as a baffle-plate or wire screen which catches floating algae, and other minute organisms, and which eventually grows large enough to change the configuration of that sea-bottom. In other words, cause and effect might be reversed.

TRANSGRESSIVE AND REGRESSIVE VIA DIFFERENT ROUTES

The application of transgressive and regressive bioherm development to this juncture has been in cross section and in its simplest form, and geologists with wide field experience could, no doubt, call attention to many objections to this theory. The writer is aware of many of these, and submits herewith an example of the difficulties which may give rise to erroneous application and faulty correlation. It is conceivable that during the advance of a transgressive sea the bioherm-forming accumulations could develop along one route (TT′) and develop via another (RR′) with the retreat or regression of the sea, as indicated in Figure 16. Such a history might indeed give rise to erroneous correlations as suggested in the cross section in Figure 14. With only subsurface cuttings and cores available, the geologist might correlate bioherms at X as the time and stratigraphic equivalent of bioherms at Y. Such a correlation would be justified in view of the fact that the underlying clastic marine sediments would appear lithologically the same and contain essentially the same deeper-sea fauna below both groups of bioherms. A thickening of the underlying dark shale would be postulated at bioherms Y, and correlation would be regarded as satisfactory from both a lithological and paleontological viewpoint for the reefs themselves and the underlying sediments. However, the difference of sediments within and above

[19] J. H. F. Umbgrove, "Coral Reefs of the East Indies," *Bull. Geol. Soc. America*, Vol. 58, No. 8 (August, 1947), pp. 732 and 772.
Ph. H. Kuenen, "Geology of Coral Reefs," *Snellius Expedition*, Vol. V, Pt. 2 (1933). Kemink En Zoon N.V., Utrecht.

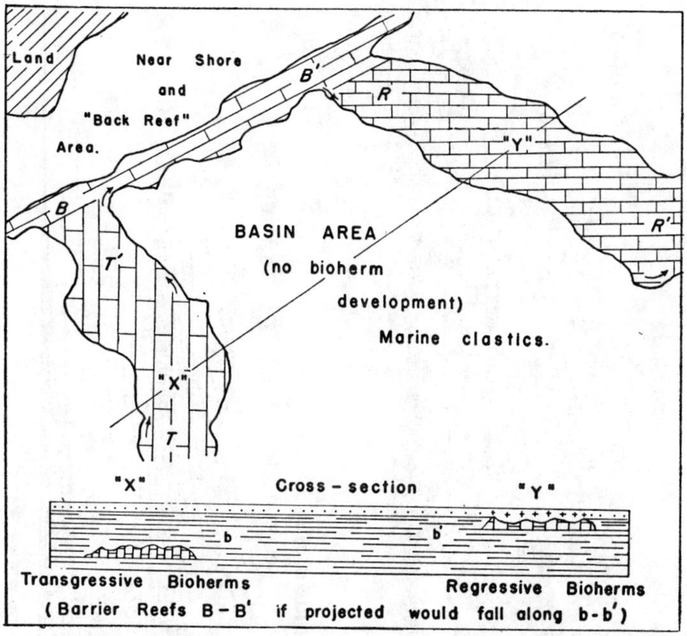

FIG. 16.—Transgressive and regressive reef deposition via different routes, showing area of transgressive bioherms (T-T'); barrier reefs (B-B'); and regressive bioherms (R-R'). First bioherms to be deposited are at T and last at R'. Cross section indicates "step-up" in stratigraphic column from reefs at "X" to those at "Y." Map is therefore a facies map with considerable time range, since deposits indicated are not contemporaneous. Section from B to T would correspond with Stage III (Fig. 4), and section from B' to R' would correspond with Stage V (Fig. 10).

the reefs should give cause for concern, since in the case of the transgressive bioherms the overlying sediments would be deeper-water clastics, and in the case of the regressive bioherms the overlying sediments might be near-shore clastics or evaporites, and within the bioherms evaporites such as anhydrite might be observed.

ADVANCE AND RETREAT OF SEA WITHOUT BIOHERM DEVELOPMENT

From the foregoing it is inferred that an advancing sea, with potential bioherm-forming organisms, could invade an area without their development into reefs or bioherms, and the same applies to a retreating epi-continental sea. Also, bioherms could develop during the advance of a sea, and none during the retreat, or *vice versa*. Such was the case in the Mackenzie River area of northwest Canada where only transgressive reefs are observed, and the Upper Devonian clastics above the reefs grade into beds of Mississippian age without evidence of regressive reefs having developed in the area observed. Attention is called to such apparent digressions from the theoretical lest the reader might regard them as contrary to the theory herewith proposed. Nevertheless it appears highly im-

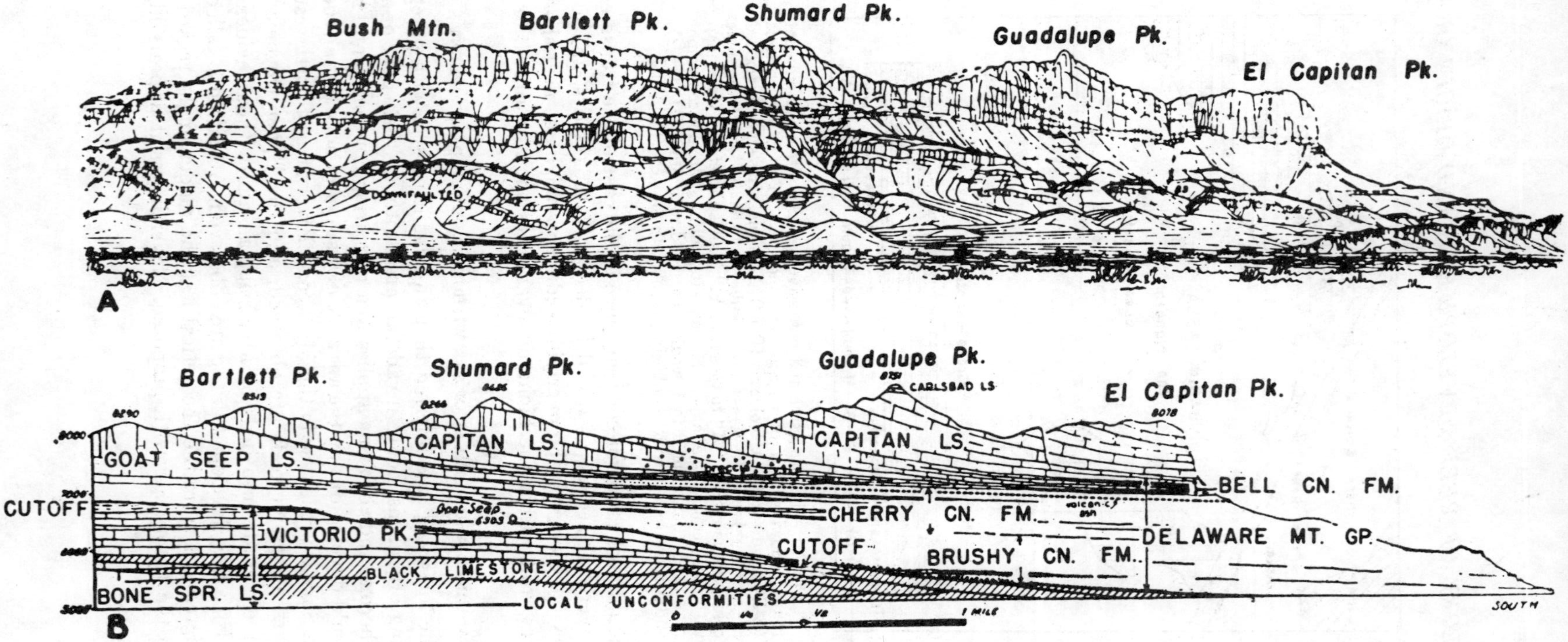

FIG. 17.—Guadalupe Peak section as interpreted by P. B. King, showing basinward regression of Capitan reefs (at right). Panorama (A) and section (B) of west side of Guadalupe Mountains.

probable that an invading or recessional sea, containing potential bioherm-forming organisms, would not somewhere encounter local conditions conducive to their growth if the area invaded were in tropical or subtropical latitude.

APPLICATION TO WEST TEXAS-NEW MEXICO PERMIAN BIOHERMS

A study of the published paleogeographic maps and history of the Permian of North America indicates that Late Permian was a time of continent-wide or over-all withdrawal of the Paleozoic epi-continental seas. This applies to North America in general, and to the West Texas-New Mexico region in particular, as attested by examination of Schuchert's well known paleogeographic maps, and more detailed maps prepared for the West Texas-New Mexico area by King and Hills.[20] More specific is the cross section by King[21] (Fig. 17), of the Guadalupe Peak area, on which either a stable strand line or the recession of the sea during deposition of the El Captian reef-limestone is strikingly indicated by the basin-ward migration of the reefs. According to Hills, Lloyd, King, Skinner, Lang, Adams, and others who have written about the Permian of the West Texas-New Mexico area, the withdrawal of the sea was not progressive and was accompanied by minor rhythmic advances and retreats of the sea, with an over-all or general withdrawal. This is what might be expected during a period such as represented by the Permian which preceded major diastrophic movements on a large scale and terminated the Paleozoic. Darwin's contention that bioherms or reefs grow only during submergence or a rise of the strand line seems to create a dilemma or paradox when applied to the growth of reefs in the West Texas-New Mexico area. If one accepts Darwin's view, the tremendous growth of reefs during El Capitan time must be explained by postulating a pronounced pulsating recession of the sea. However, this is not of necessity the only explanation, since a basin-ward growth of the El Capitan reefs is clearly in evidence and is so well illustrated in King's section at Guadalupe Peak. From this it appears that reef or bioherm development can take place during an over-all, gradual lowering of the strand line on a large scale, without extreme oscillations. The writer is inclined to believe that during the El Capitan reef growth the strand line dropped fairly slowly, and in that local area the main regression set in at the end of this delta-like reef deposition. Furthermore, the reefs developed on the platform east of the Delaware basin might not be the exact time equivalents of the El Capitan reef.

SUMMARY OF SALIENT FEATURES OF TRANSGRESSIVE AND REGRESSIVE BIOHERM DEVELOPMENT

The main characteristics of the *regressive* type of reef or bioherm deposition such as the El Capitan type are, according to Lloyd and others, as follows.

[20] J. M. Hills, "Rhythm of Permian Seas," *Bull. Amer. Assoc. Petrol. Geol.*, Vol. 26 (1942), pp. 217–55.

[21] P. B. King, *op. cit.*

1. Back-reef evaporites, clastics, and brecciation
2. Massive, porous reef, or bioherm accumulation
3. Fore-reef, well bedded clastics such as shale, limestone, sandy limestones and dolomite, with occasional wave-developed or slumped breccia
4. Basin or pontic well bedded gray to black shale, sandy shale, and limestone

Facies 1 and 2 overlap 3 and 4 during a normal retreat of the sea. The regressive type of bioherm or reef may contain within it back-reef types of sediments such as red shale and anhydrite, but the transgressive type does not.

The transgressive type of reef or bioherm deposition lacks the typical back-reef facies, so that the shoreward and basinward deposition, and the later overlying sediments are essentially the same. In many places there are well bedded basin-type shales surrounding the bioherms completely, as at Norman Wells; in other places, such as the smaller bioherms in southwest Ontario, the surrounding sediments are limestones.

Both the transgressive and regressive types could give rise to blanket or compound reefs extending laterally for many miles, depending on whether or not the transgression or regression was pulsating. In fact they might develop into biostromes instead, as seems to be the case of the D-2 zone at Leduc, Alberta. Examination of the geologic column for the Leduc area brings this out in Figure 5 which shows the D-2 zone of fairly uniform thickness, overlain by evaporites and draped over the D-3 zone. It is therefore conceivable that, with extreme pulsation during both transgression and regression, the results might be a hybrid type of reef and facies development which could be deciphered only after many data were made available. Destruction of the upper part of a transgressive bioherm by wave action and sub-aerial erosion is less likely than in the case of the regressive type, but wave-base or submarine erosion is possible and probable.

For sake of emphasis, it is repeated that, due to the environment and conditions under which a transgressive-type reef develops, the absence of evaporites and other related back-reef types of sediments within the bioherm and surrounding it is a necessary prerequisite. Both the Norman Wells reef and the D-3 zone at Leduc are surrounded by the basin-type sedimentary facies, thus suggesting a different history of their growth as compared with the late Permian El Capitan bioherms. This does not preclude the possibility, and indeed the probability, that transgressive types of reefs have been, or will be, found in the Permian during that phase of Permian time when the sea was advancing over the area, as suggested in Figure 15. In other words, there may be transgressive reefs in the Delaware basin, for the simple reason that the bioherm-forming organisms of the El Capitan reef did not fly in; they had to come in by water with the advancing sea (Fig. 17). The Victorio limestone at the lower left in the section, overlain by clastics, could be, or better still some earlier Permian or late Pennsylvanian reefs might be the transgressive type of reefs which brought in the organic assemblage which gave rise to the El Capitan reef. However, as pointed out, the transgressive Permian sea could have invaded the area without reef developments, although potential reef-forming organisms were present.

CONCLUSIONS AND IMPLICATIONS

In conclusion, it may be stated that the conception of transgressive and regressive bioherm or coral-reef development is indeed not a new departure from geologic thinking, and it is very simple indeed. The implications of these concepts are as obvious as the concepts are simple. One application is herewith submitted as an example.

Let us suppose a wildcat well is being drilled in virgin territory, and let us suppose that the cuttings and cores reveal the presence of evaporites overlying or associated with coral-reef or reef-like porosity containing either gas, oil, or water, or some of each. In other words, there has been established the presence of a regressive type of bioherm or biostrome. According to the reasoning presented, there could be below this wildcat, or somewhere in this area stratigraphically below the regressive reef discovered, a transgressive type of bioherm or biostrome, for the simple reason, as already stated, these bioherm-forming organisms did not fly into this area; they had to come in with the advancing or transgressive sea. Obviously that wildcat test should be carried deeper, or others should be drilled deeper in that area with the hope of finding the transgressive bioherms lower in the section. However, the transgressive and regressive routes may not have been the same, and thus the deeper drilling in this case might eventually discover the transgressive bioherms in a near-by area, rather than directly beneath the regressive bioherms. Nevertheless, the principles still hold, and other implications become obvious. A regressive reef overlying a transgressive type should suggest a merging of the two shoreward into a combination of the two, giving rise to a thick barrier reef *et cetera*. The beauty of this sort of geological thinking is that there seems to be no limit to one's imagination as more and more data become available and thus more vistas are opened for speculation. Deviations form the idealized conditions as herewith presented may therefore be reconciled with the theory only when sufficient data have been gathered and correctly interpreted.

PART II

ORIGIN OF HYDROCARBONS AND POROSITY WITHIN BIOHERMS

In the presentation of this paper on transgressive and regressive development of bioherms at the St. Louis meeting, 1949, the question of the origin of hydrocarbons within them, as well as the porosity, was mentioned.Further discussion of these topics is presented herewith. It involves the unsolved problem of the origin and migration of oil or hydrocarbons—a problem beset with geological prejudices and preconceived notions which have permeated our thinking to a marked degree. The writer is submitting this contribution merely as food for thought along these lines, and not as established or finalized deductions.

HYDROCARBONS INDIGENOUS TO CORAL-REEF OR BIOHERM RESERVOIRS

The majority of petroleum geologists[22] think of the surrounding shale and the underlying sediments as the source rock of the oil accumulated in the porous parts of coral reefs or bioherms. Others favor the idea that the hydrocarbons were not only reservoired in the bioherm, but were also generated within the reef itself. When one considers that in coral reefs or bioherms conditions exist, ideal for the growth, death, and accumulation of countless generations of innumerable organisms, why is it necessary to look for outside sources to explain the oil found within them?

According to Pettijohn,[23] "The bioherms were islands of intensive vital activity" and all observations made and recorded by students of modern coral reefs bear this out. If the organic theory of the origin of oil is adopted it is reasonable to assume that a porous source rock would also serve as the reservoir rock, as might be the case with bioherms or coral reefs, as well as biostromes, or any other porous limestones or dolomites. There are geologists who entertain the idea that sandstones could have hydrocarbons generated within them and thus act as both source and reservoir rock.

It should also be remembered that in addition to the lime-secreting organisms of a reef or bioherm, there are probably more of the type which do not leave behind them any hard parts, and the organic remains of these merely add "more fuel to the fire." The accumulated soft parts of the planktonic types may be greater than those of the sessile types within a bioherm.

The concept of an environment wherein organisms would die because of exceptional or catastrophic conditions, and thus wipe out an ecological assemblage to account for the generation of hydrocarbons, does not appeal to the writer nearly so much as the more plausible concept of an environment conducive to the natural growth of a fauna and/or flora, so that generation after generation would live and die by natural causes and accumulate almost *in situ*. There are also many arguments against this idea, and one should maintain an open mind; nevertheless, recent chemical studies by Bergmann and Lester[24] indicate that modern "corals contain minute amounts of a wax-like substance which apparently becomes entrapped in the ever-growing reefs. The wax, which makes up about one seventh of one per cent of the coral, consists largely of hydrocarbons and complex alcohols, chemicals similar to those in petroleum."

Interpretation or deductions from observed field data are commonly influenced by pre-conceived notions or faulty geologic prejudices, and thus data may be presented in favor of a plausible interpretation which is diametrically

[22] J. E. Adams, "Origin, Migration and Accumulation of Petroleum in Limestone Reservoirs in the Western United States and Canada," *Problems of Petroleum Geology*, Amer. Assoc. Petrol. Geol. (1934), pp. 347–63.

[23] F. J. Pettijohn, *Sedimentary Rocks*, p. 298. Harper and Brothers, New York (1949).

[24] W. Bergmann and D. Lester, "Coral-Reefs and the Formation of Petroleum," *Science*, Vol. 92, No. 2394 (November, 1940), pp. 452–53; *Science News Letter*, Vol. 54, No. 18 (1948), p. 281.

opposite to the true or correct deduction. An example of this would be the presentation of the outcrop section Figure 17 at El Capitan prepared by King, or a similar one from elsewhere, in an attempt to demonstrate that the so-called pontic or basin facies dark shale and limestones are the source rock and that the fore-reef clastics are the carrier beds because the latter still retain some oil in them at the outcrop and because the basin facies are black bituminous shale and limestones (Fig. 18).

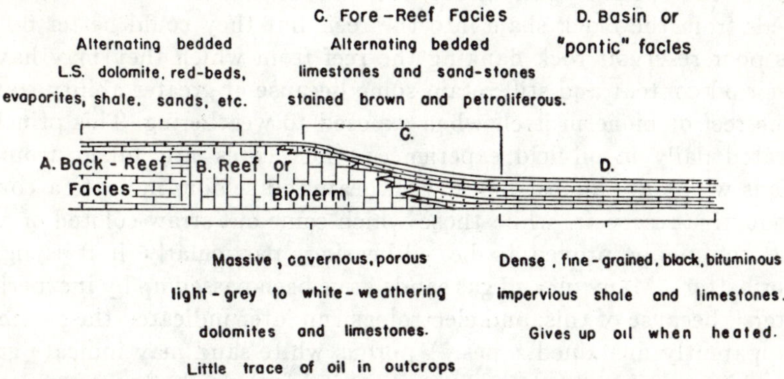

FIG. 18.—Ideal outcrop diagram of back-reef (A), reef or bioherm (B), fore-reef (C), and basin (D) sediments which may be interpreted in two ways as follows.

Interpretation A.—Basin or pontic facies (D) is source rock; fore-reef facies (C) are carrier beds; reef rock (B) is reservoir rock of oil which migrated from D through C to B.

Interpretation B.—Basin or pontic facies (D) are source rock which still retain their original bituminous material; fore-reef facies (C) are combination of source and reservoir rock; bioherm or reef rock (B) is main source rock as well as original reservoir rock.

Such possible erroneous deductions can best be dealt with by considering the sediments in question and their reaction to weathering. A porous medium such as limestone or dolomite from within a bioherm, which obviously contained oil at one time will, when exposed to rain, sunshine, and the elements lose practically all of its hydrocarbons in a very short time, excepting such oil as may be entrapped in vugs or pores which are not connected with the surroundings. In such cases the oil eventually turns into tar or solid black bitumen. As a result the reef limestones such as the El Capitan will appear in outcrop as almost white or gray, and show very little signs of ever having contained oil and/or gas.

A medium- to coarse-grained sand loses the light hydrocarbons less readily, a fine-grained sand or silt retains them longer, while a fine-grained limestone or shale may retain them almost indefinitely or at least until broken up, pulverized, and heated. The oil in such shales or dense limestones as the pontic facies of El Capitan time or the Fort Creek shale in the Mackenzie River area can not move out of them excepting through fractures in limited amounts. In other words, the hydrocarbons within such dense limestone or shale bodies are indigenous to them and will remain in them almost indefinitely, unless subjected in nature to a process comparable with that employed to-day in attempting to extract these hydrocarbons commercially. Mountain-making forces and temperatures might accomplish this. Such world-wide black shales of various geologic ages contain indigenous hydrocarbons which they may never give up, and because of their fine texture will not allow free migration from them excepting through fractures.

The same applies to siltstones or fine-grained sandstones to a less degree. Therefore, in the writer's opinion, the fore-reef facies are not to be regarded as carrier beds from the black shale into the reef, but they could better be interpreted as poor reservoir rock flanking the reef from which they may have obtained their oil content and still retain some because of greater ability to retain it than the reef or bioherm itself when exposed to weathering. This principle is demonstrated daily in oil-field experiences when cores are taken from drill holes. Cores which are almost black with heavy oil seldom indicate a commercially productive reservoir, while those which come out straw-colored or almost white are commonly proved to be oil-bearing, particularly if it is light oil (20° Bé. or better). Many oil and gas sands have been passed up by inexperienced "well sitters" because of this, and electrologs run later indicated the possibilities of such apparently unstained zones. A porous white sand may indicate gas, oil, or water, while a uniformly stained core may indicate dead oil with no pressure.

Objections to foregoing reasoning are numerous, and the one favored is the suggestion that the oil migrated through what is now black shale at the time it was a mud, ooze, or silt. That leads to the all-important question "Just what is the nature of virgin oil?" Years ago we were taught that it was a heavy, tarry, or asphaltic substance, but this idea has lost favor, and many, including the writer, believe it may be a very light substance. If it is a heavy, tarry material, its migration through the muds is a serious physical problem from the outset. If it is a light substance and generated before consolidation of the sediments it would have a tendency to move upward or vertically through the muds and oozes, and from these to the surface of the water and thus be dissipated. Gas is one of the first manifestations of hydrocarbons "in the making" as attested by observations of swamps, sloughs, ponds, or marshy embayments of larger bodies of water. The writer believes that the presence of marsh gas may be the first step in the formation of hydrocarbons. Therefore, if a virgin oil is a light substance, its migration from muds or oozes would be primarily upward or vertically, and the wholesale or widespread migration of oil through carrier beds would require some lithification of the sediments before it can take place. However, if this rock is too fine-grained and dense, no appreciable movement of oil through it seems possible.

Expressed in another way, the writer regards the fine-grained black shales observed throughout the geologic column as source beds which retained their original oil because of its inability to migrate through such sediments after consolidation. It could be termed "still-born oil," and its movement out of such im-

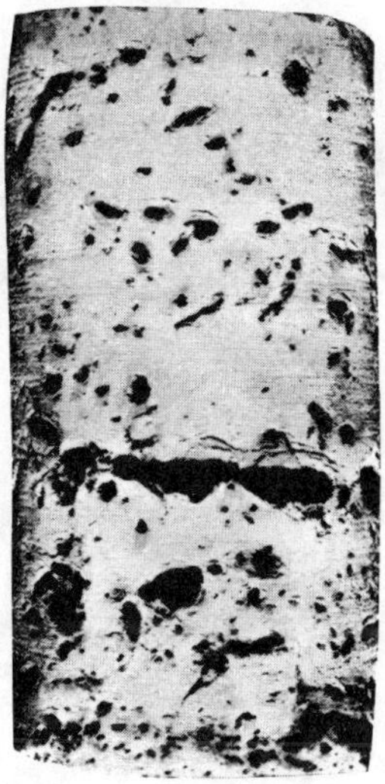

A

B

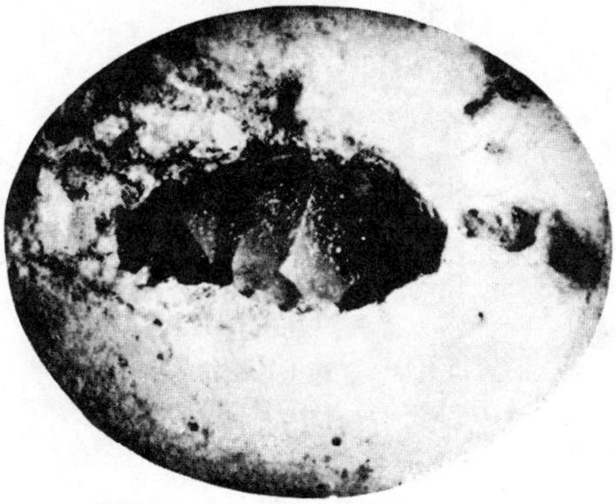

C

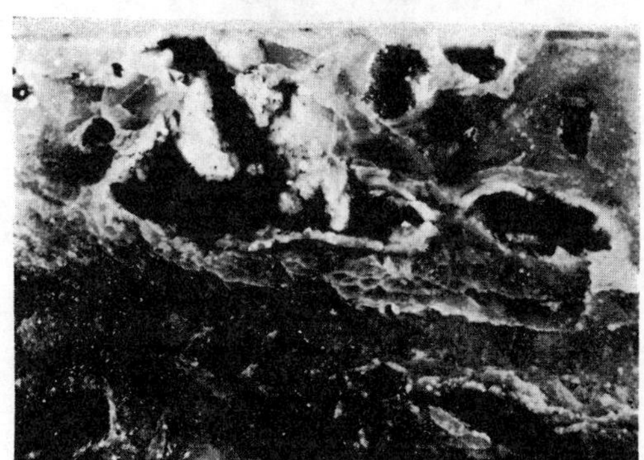

D

PLATE I

FIG. A.—Core from D3 zone, Imperial's Redwater No. 3, illustrating very porous bicherm (reef) limestone in which porosity is due to hollow interior of hard parts of organisms and voids due to their helter-skelter accumulation. (Diameter of core $3\frac{3}{4}$ inches.)

FIG. B.—Same as Fig. A with fractures which increase effective porosity and vertical permeability. (Diameter of core $3\frac{3}{4}$ inches.)

FIG. C.—Part of Imperial's Redwater No. 3 core (D3 zone) illustrating primary pore or cavity in which secondary calcite crystals were precipitated. (Cavity approximately $\frac{1}{2}$ inch in diameter.)

FIG. D.—Part of core from Leduc, Alberta D3 bioherm (reef) dolomite, illustrating precipitation of secondary dolomite crystals on walls of primary pores or cavities. (Width of illustration approximately $3\frac{3}{4}$ inches.)

Note.—All illustrations on this plate were supplied by D. B. Layer through courtesy of Imperial Oil Limited, exploration department 4, Calgary, Alberta.

A

B

C

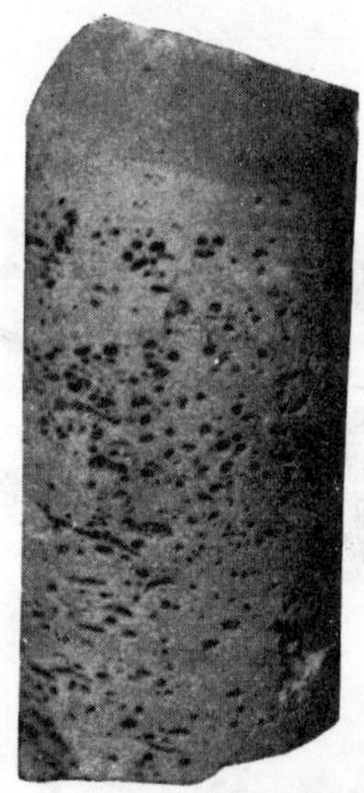

D

PLATE II

FIG. A.—Polished cross section of dolomite bioherm (reef) core taken from D3 zone, Imperial's Leduc No. 43. Primary porosity due to hollow interior of organisms and sugary or crystalline dolomite matrix. Depositional banding and minute fractures, increasing effective porosity and permeability. (Diameter of core 3¾ inches.)

FIG. B.—Same as A, with fractures more apparent and local concentration in layer or banding of hard parts of organism near top of core. (Diameter of core 3¾ inches.)

FIG. C.—Core of D1 zone, Leduc, Alberta, showing dolomite breccia, probably due to solution of evaporites within and below, causing collapse of partly consolidated sediments. Dark angular fragments are oil-stained dolomite. (Diameter of core 3 inches).

FIG. D.—Core of D2 bioherm or biostrome dolomite zone, Leduc, Alberta. Porosity due primarily to interior of corallites. Such porosity is not most effective; permeability, if not accompanied b fracturing, is low. (Diameter of core 3¾ inches.)

Note.—All illustrations on this plate were supplied by D. B. Layer through courtesy of Imperial Oil Limited, exploration department, Calgary, Alberta.

pervious sediments could be effected only when subjected to intense deformation, high temperatures, and fracturing. A coarse-grained or porous rock above, below, or within such black shale could eventually accumulate some of the hydrocarbons from the shales, but the fact that the shales are black does not necessarily imply that at one time they contained much more oil. Such black shales are both the source and the reservoir rock which may never give up its hydrocarbons, excepting as here described.

POROSITY OF BIOHERM RESERVOIRS

Coral-reef or bioherm limestones or dolomites can best be described as a homogenous heterogeneity of porosity which confounds the mathematical approach of the petroleum engineer to calculate their gas and oil reserves, and which likewise baffle the geologist as to size and probable configuration, for the simple reason that no two bioherm reservoirs are identical because of their inherent complex manner of accumulation and growth (Pls. I and II)

In addition to the irregular and unpredictable pore space within a bioherm it is commonly criss-crossed by minor adjustment faults and fractures which develop soon after the bioherm becomes large enough to begin settling and adjusting itself to its surroundings, and later to its overburden. Because of these factors the accurate estimate of the ultimate recoverable reserves of a coral-reef (bioherm) reservoir can not be calculated early in the history of such an oil pool, and rarely as accurately as those for sand-reservoir pools. Not until a production history has been obtained can fairly accurate reserve estimates be made.

The porosity in true bioherms such as the D-3 zone at Leduc is regarded as primary in origin. The pores came into existence at the time of deposition and growth of the bioherm, and are not regarded as solution cavities. The pores are not in all instances regarded as the hollow interior of corallites *et cetera*, but represent what remains of the original open spaces between the individual lime-secreting organisms of the bioherm. Initial or primary porosity of a bioherm can best be described by comparing it with a strawstack which has been compressed by an overburden. The individual hollow pieces of straw have their porosity, and the helter-skelter stacking of the straw gives rise to even greater pore spaces. The same thing takes place in a bioherm which dies and is buried by later sediments. A sizeable percentage of the original empty space between the variously shaped organisms is retained, and thus gives rise to the odd-shaped pores found in such a bioherm as the D-3 zone at Leduc, Alberta. As stated, some of these can definitely be interpreted as the hollow insides of corallites *et cetera*, but most of them are the reduced empty spaces between the broken-down corals, algae, sponges and bryozoans, and this type of porosity is the most effective, because of continuity. Since there are changes of conditions during the growth of a bioherm, porosity variations are to be expected, and zones or layers of variable porosity or density are inevitable.

The presence of dolomite linings within pore spaces does not indicate that they are solution cavities. These dolomite or calcite linings are incipient secondary deposition or crystallization which, if continued, might eventually obliterate all the porosity. However, such could not have taken place on a large scale in the D-3 zone in the Leduc area because of its complete sealing-off by the green shale which surrounds and overlies it. The D-2 zone presents a different condition. This bioherm (or biostrome) is overlain by anhydrite and evaporites. Waters charged with anhydrite, calcite, and dolomite entered the D-2 zone and partly obliterated its primary porosity by secondary deposition on these minerals. Observation of cores indicates varying degrees of such secondary crystallization or obliteration of porosity. In some places in the Leduc field it is complete and no production of oil, gas, or water is obtained.

The overworked term "solution cavities" is inherited from the phase of the West Texas Permian development when geologists, unaware of, or unwilling to accept, the coral-reef reservoir conception, were attempting to explain the porosity observed or inferred from cable-tool cutting as that developed in limestone, or dolomite by solution. However, when such a bioherm like the El Capitan reef is eventually raised near enough to the surface to allow circulating ground waters to enter, solution begins, cavities develop, grow larger, and result in such caves as those at Carlsbad, New Mexico. If, by chance such a chambered bioherm is later buried, it could act as a reservoir for the accumulation of oil and/or gas, which is not indigenous, but derived from some other reservoir at depth nearby if a pressure-gradient exists between the two. Obviously, such an accumulation would be expected at or close to an unconformity, as is the case in parts of the Kevin-Sunburst field of Montana where oil is produced from large cave-like chambers in the Madison limestone at the Jurassic-Mississippian unconformity.

CORAL-REEF OR BIOHERM BIBLIOGRAPHY

In this contribution the writer has made no attempt to refer to all the excellent publications relevant to the subject of coral reefs which have appeared since Darwin's time and previously. To name some of the contributors would do an injustice to those who would be omitted. In recent Bulletins of the Geological Society of America many contributors on coral-reef study have appeared as the result of the Bikini project and American occupation of the Pacific coral-reef islands. Previous to that time the scientific literature of many countries contained numerous excellent contributions on the subject. Grabau's *Principles of Stratigraphy* contains excellent summaries and diagrams of ancient or fossil coral-reef studies. The West Texas and New Mexico reefs of the Permian can be studied in the pages of our own A.A.P.G. *Bulletin*. Nevertheless the writer believes that one of the first "musts" for coral-reef or bioherm study is the reading of Darwin's classic descriptions and logical deductions. The writer feels certain that if Darwin had seen the El Capitan reef outcrops he would have concluded that reefs also develop during a lowering of the strand line. The writer would like to make the suggestion here that the compilation of a comprehensive bibliography on coral-reef or bioherm literature would be an excellent A.A.P.G. or G.S.A. project which might be dedicated to the memory of Charles Darwin whose power of observation, description, deduction, and clear-cut writings deflected the writer of this contribution into the earth sciences which have proved to be so fascinating.

Added Discussion[25]

The presentation of this contribution on the pre-Christmas (1949) portion of the lecture tour[26] brought forth considerable questioning from the various audiences, and this added discussion is intended to cover the most frequently raised points as follows.

(a) Why is the D-3 zone in the Leduc field *dolomite*, while at Redwater it is *limestone*, even though both are surrounded and overlain by the same or equivalent "green shale" of the Upper Devonian?

The writer believes that conditions or the environment during or soon after accumulation of the reefs or bioherms is in part the answer. Just what this difference in conditions (chemical or physical) may have been, may be ascertained from the study of recent reefs which (according to observers) indicate dolomitization in some areas during or very soon after accumulation. Recent drilling through recent reefs also indicates progressive dolomitization at depths as well as alternating zones of dolomite and limestone. The "back-reef" facies in the Permian of New Mexico also shows alternating layers of limestone, red shale, dolomite, anhydrite, *et cetera*, suggesting that slight change in conditions gave rise to different types of deposits in the same locality.

(b) The observed saucer-shape of the bottom of many reefs may be due to the more rigid reef settling into the underlying and surrounding shale, but the writer believes it may be due primarily to progressive upward and outward growth of a bioherm, contemporaneously with the less rapid accumulation of the surrounding clastic sediments, just as an individual coral colony takes on globular, spherical, or ellipsoidal shape. In other words, one coral colony may be the beginning of a bioherm which grows upwardly or vertically, as well as laterally and which, under ideal conditions, would assume spherical shape. From well cuttings and lithologic character such a condition could very easily be misinterpreted as structural rather than stratigraphic, just as is the case with the top of a reef or bioherm.

(c) *Regressive reefs* or bioherms can, and no doubt do, develop *without back-reef evaporites et cetera*, if they are not of the barrier type, since under such conditions no shallow lagoons would develop behind them shoreward. However, such isolated regressive reefs would probably be flanked on the shoreward side and overlain by coarser clastics than on the basinward side. In other words they would be overlain by regressive sandstones *et cetera*. It is also possible that a regressive type of reef could be covered by some shale or dense limestone if, after it ceased growing, the strand line remained stationary long enough to allow such clastics to accumulate to depths as great as, or greater than, the vertical height of the reef. This is intimated or inferred in the original text.

(d) *Shale and dense limestone partings* within a reef or bioherm merely indicate a local or temporary cessation of growth of the reef, which covering of shale, in some instances, could have been complete over the entire reef. "Re-seeding" by fertilized ova from nearby would cause resumption of reef growth over the shale or dense limestone layer if all other factors for reef growth are satisfied. Tongues or wedges of such shale or dense limestone extending into the reef are the rule rather than the exception.

[25] Manuscript received, January 2, 1950.
[26] Tour arranged by A.A.P.G. distinguished lecture committee.

(e) *The great thickness of the El Capitan group of reefs* of West Texas and New Mexico could be due to an originally steeper slope at the site of accumulation, or as suggested by John Emery Adams, to a contemporaneous down-warping in that area, or a combination of both. All these possibilities are discussed by Philip B. King in his contributions already referred to in the text.

(f) Both the Leduc and Redwater fields were discovered on the basis of seismic-reflection surveys, and future discoveries will no doubt depend on improvements in seismic technique and interpretations combined with structural and stratigraphic principles properly applied, as well as imagination on part of both geologists and geophysicists.

(g) In speaking of a *rise in the strand line* the writer does not wish that phrase to be taken literally any more than the incorrect reference to the "rising" or "setting" sun. We know that the sun does not *rise*, nor does it *set*, but that the earth rotates, thus making it appear that the sun rises and sets. A rising strand line, or encroachment of an epi-continental sea, may be due to a down-warping or subsidence of an area, with no change in sea-level. As a matter of fact that is probably the way in which most epi-continental sea invasions came about.

Daly's "glacial control theory," to which already too much space has been allotted in the literature, has never been taken too seriously by the writer for the simple reason that the few feet of rise or drop in sea-level due to melting or growth of the ice caps is too small to account for the hundreds or thousands of feet of coral-reef growth observed in present-day reefs. Differential subsidence or down-and-up-warping can readily explain most of them. In this connection, the writer refers to his first published and embryonic contribution to the science of geology entitled "Post-Tertiary Strand-Line Oscillations in the Caribbean Coastal Area of Colombia, S.A.,"[27] which, strangely enough, touched upon "coral reefs" studied and observed by him in the Caribbean Sea 27 years ago, and in which paper Daly's glacial control theory was regarded as unnecessary.

If the melting of the Pleistocene ice caps caused the rises of the strand line, and that in turn caused coral-reef growth, how about postulating the melting of ice caps for all past geologic periods to account for their reef or bioherm development?

[27] *Jour. Geol.*, Vol. 35, No. 1 (1927), pp. 58–72.

Reprinted from
BULLETIN OF THE AMERICAN ASSOCIATION OF PETROLEUM GEOLOGISTS
VOL. 36, NO. 11 (NOVEMBER, 1952), PP. 2125-2149, 4 FIGS.

FACIES RELATIONSHIPS OF ORGANIC REEFS[1]

PRESTON E. CLOUD, JR.[2]
Washington, D. C.

ABSTRACT

Organic reefs are defined in terms of potential to create a wave-resistant structure. The term "bioherm" may be retained for ancient reef-like organic masses of uncertain potential or doubtfully wave-resistant nature. The frames of organic reefs are built by a variety of organisms, and the reef mass includes widely varying but commonly large proportions of interstitial clastics.

Post-Paleozoic reefs built by corals and algae presumably grew at depths less than 25 fathoms and in a tropical or subtropical environment. In the case of reefs more than 150 feet thick this implies relative subsidence. For Paleozoic reefs the same general order of magnitude of depth and temperature is believed probable, especially if algae were abundant. Without regard to time, calcareous algae as abundant contributors to reef formation strongly indicate depths less than 50 fathoms and probably less than 25 fathoms. More specific limits of depth are implied by particular families of algae ranging back to Early Ordovician. The presence of calcareous green algae is presumptive evidence of a tropical or subtropical environment.

Subequidimensional or irregular organic structures known as patch reefs suggest growth in relatively quiet waters. Such structures generally grade abruptly to adjacent sediments. Table reefs are flat-topped, isolated reef masses of the open ocean, without true lagoons and with or without islands. Some resemble large patch reefs, but others were probably deriv ed from atolls.

Linear reefs—fringing reefs, barrier reefs, and the ring-like outer reefs of atolls—are typically wave-fronting structures. The growth forms of their wave-facing elements are characteristically crustose, stubbily branching, or massive. Seaward the linear reefs grade through reef talus to enveloping sediments. The nature of the contact is related to source and sedimentary environment of offshore deposits and to slope of reef front. Fringing reefs and their fore-reef talus should characteristically show abrupt contacts with the land surface which they veneer, and both the reef proper and its talus may contain variable amounts of land-derived detritus. Barrier reefs and the outer reefs of atolls should show complex relationships with lagoonal sediments.

The Middle Devonian patch reefs of Michigan are analogous to those of existing seas, and the Middle Silurian reefs of the Great Lakes region were wave-breaking or potentially wave-resisting structures. In the Permian Capitan barrier reef the frame-building organisms presumably were sponges and calcareous algae, but physico-chemical precipitation of calcium carbonate probably contributed to the reef mass. Massive Cambrian bioherms in central Texas appear to have resulted almost entirely from the calcium-precipitating and sediment-binding activities of blue-green algae. Some *Scolithus* sandstones may be analogous to the annelid reefs of the North Sea. Other structures called reefs or bioherms may be depression fillings or banks of triturated organic materials that were held together, as they piled up, by concordant deposition of surrounding sediments.

INTRODUCTION

New information about organically constructed reefs and their foundations is accumulating rapidly and in volume. There has been, since 1945, an unprecedented amount of drilling into modern reefs; studies of their surface features and ecology; deep dredging; and geophysical investigation of atolls, seamounts, and their environs. In addition, the growing awareness of the importance of ancient reefs and reef-like bodies in the localization of oil has led to increased interest in their structure, lithic nature, and paleontological peculiarities.

Under such circumstances it is desirable periodically to focus attention and discussion on critical or poorly known elements of the reef problem, and a number of excellent summaries have recently been published. Generalizations about

[1] Manuscript received, July 28, 1952. Publication authorized by the director of the United States Geological Survey.

[2] Chief, Paleontology and Stratigraphy Branch, United States Geological Survey.

facies relationships of reefs without special reference to time are yet few, however, and are faced with unusual difficulties. Paleozoic and post-Paleozoic reef-building corals belong to different subclasses of the Anthozoa and show significant differences as well as similarities in their growth habits and associations. It is uncertain exactly what conditions controlled the growth of certain now essentially structureless reef-like masses in the older rocks. Many of the best known existing reefs are surrounded by water deeper than that which surrounded most known reefs of the ancient epicontinental seas. The association of reefs with evaporite deposits is unknown in present seas.

Yet it is concordant with known evidence of other sorts to consider that Tertiary reef-building organisms were affected by controls approximately identical with those imposed on existing representatives of the same species, genera, or families, and only a little less certain that the Mesozoic reef-builders underwent similar restrictions. Moreover, some common Paleozoic reef-dwelling algae belong to widespread and ecologically distinctive living families, and the mere presence in reefs of abundant algal structures presupposes water within the range of vigorous photosynthetic activity. It is also reasonable to infer that Paleozoic reef-building corals were probably limited to depths and temperatures of approximately the same order of magnitude as are existing corals of similar association and habit, for high calcium metabolism in itself appears to be characteristic of relatively shallow tropical and subtropical waters. Above all, the basic principles that affect sorting and arrangement of particles, growth form, and relative solubility of chemical elements do not vary with time. And, finally, there are existing reef tracts that may be studied with profit in the search for analogies with ancient reefs—namely, those of Indonesia, the Philippines, and the shelf seas of Asia proper, Australia, and New Guinea (in part known from the work of Umbgrove, Kuenen, Krempf, Fairbridge, members of the Great Barrier Reef expedition, and others).

In the ensuing paragraphs an attempt is made to summarize some significant known relationships of existing reefs and to discuss briefly some features to be looked for in the broader facies relationships of reefs, wherever they may appear in the geologic column. This is done on the basis of admittedly incomplete data and in full awareness that some of the generalizations made are so sweeping that detailed studies in progress, or to be made, may prove them in part invalid. Whether or not the suggested generalizations are substantiated, the effort will have been in order if they help to bring the rapidly accruing data into better focus or elicit critical comparisons with known reef structures.

Kinds of Reefs

It is desirable to define the kinds of organic reefs and discuss their descriptive terminology before considering their facies relationships. Organic reefs are or were actually or potentially wave-resistant mounds, platforms, or linear or irregular masses that were constructed under organic influence and they rise or rose significantly above the sea floor. Where the nature of the structure referred to is clear from context, it may be called simply a "reef" without serious damage to that conventionally nautical term.

Thus, allowing for the part that may be played by physico-chemical precipitation on the reef flat or within the reef mass, the writer concurs with Lowenstam's definition (1950, p. 430) of organic reefs "in terms of fundamental biologic potentials of the organisms responsible for them rather than in terms of the present day appearance of the structures." These potentials, as given by Lowenstam "are the ability to erect rigid topographic structures by frame-building, sediment retention and binding and to create a wave-resistant structure." It is subsequently noted that the configurations, as well as the ecologic and sedimentary relationships of organic reefs, are strongly affected by their situation with reference to wave action and other factors of their physical environment (see also Ladd, 1944). However, the fact that structures now or finally wave-breaking may for much of their existence not have been so should not debar application of the term "reef" to the structure as a whole, or to similar structures that are or were only potentially wave-breaking.

The classic and conspicuous types of existing organic reef complexes are fringing reefs, barrier reefs, and atolls (Darwin, 1889, p. 1; and earlier). But there are and were equally distinctive, though less conspicuous, types of reefs.

Faros (Gardiner, 1931, p. 19) are small ring reefs that occur as lagoon reefs or parts of composite atolls, as in Mahlos Mahdulu and Nilandu atolls in the Maldives (Darwin, 1889, Pl. 2; Gardiner, 1931, Figs. 5, 7). A more miscellaneous group of reef structures comprise the ordinarily small mound-like masses, platforms, or pinnacles of generally subcircular to irregular plan that grow upward from the floors of lagoons, shallow benches, or shelf areas. Such reefs are here collectively referred to as *patch reefs* (Umbgrove, 1929, p. 34; 1946; 1947, p. 757; 1949, p. 7), whether ancient or present-day. Various types of patch reefs have also been called pinnacles (Gardiner, 1931, p. 136), chapeirões (with flaring tops, Hartt, 1870, pp. 199–213; singular—chapeiraõ), Jesuits (Hartt, 1870, p. 199), coral knolls (Darwin, 1889, p. 91), reef knolls (Umbgrove, 1947, p. 756), platform reefs (Fairbridge, 1950, p. 345), and so on. The term "*bioherm*," originally proposed by Cumings and Shrock (1928, p. 599; also Cumings, 1932, p. 333) as a substitute for the general term "reef," may be reserved for reef-like organic masses of uncertain potential or doubtfully wave-resistant nature. "*Biostrome*" (Cumings, 1932, p. 334) has been applied descriptively to tabular or thinly lenticular organic masses that are reefs in an incipient sense, as well as to shell beds and other types of structures that in no sense are to be considered reefs. And, finally, the term "*table reef*" (Tayama, 1935) has been employed, especially in discussion of existing reefs, for reference to flat-topped, isolated reef mounds of the open ocean, without lagoons and with or without islands. It must be noted, however, that some examples of elevated table reefs cited by Tayama (in the Mariana Islands) are not comparable with the existing structures to which he applied the name.

Several descriptive terms for lesser reef features are finding increasing usage in the reef student's vocabulary. For instance, coral heads with high rims and dead centers are designated *micro-atolls* (Krempf, 1927, pp. 13–19). Shikkan, plaatriffe, platform reef, shoal reef, and a number of other terms have also been applied to variants of one or another of the major reef categories by students of ancient and existing reefs. Except for those that designate the principal reef types, however, the terms used seem of little importance as long as they say what a particular kind of reef is like in fairly ordinary words. Thus, expressions that indicate the general shape, place of formation, and organic composition of a reef or type of reef are here used freely and informally.

The important terms are: *organic reef*, or *reef*, as a general term; *fringing reef* for veneering reefs that lie or lay adjacent to pre-existing land; *barrier reef* for sublinear structures that are or were separated from near-by older land by a lagoon; *atoll* for composite structures with ring-like outer reefs that surround or once surrounded a central lagoon devoid of pre-existing land; *patch reef* for small, subequidimensional or irregularly shaped reefs that are typically parts of reef-complexes; and *table reef* for flat-topped, isolated, characteristically small reef-mounds of the open ocean. Tabular or thinly lenticular organic masses composed of frame-building organisms that elsewhere build recognizable reefs are perhaps best referred to as *incipient reefs* or by some simple descriptive phrase that will distinguish them from shell beds or other *biostromes*. An ancient reef may have passed through more than one of the foregoing categories at different stages of its growth. Ancient reef-like organic masses of uncertain potential or doubtfully wave-resistant nature may be called *bioherms*.

The foregoing terms by custom refer primarily to structures formed in a marine environment, but any that are appropriate should be considered equally available for fresh-water structures.

Contributing Organisms

Although coral reefs appear to be the rule among existing organic reefs, their rule is by no means universal to-day, and many important reef structures of the past were built with little or no help from corals. It is also well to keep in mind that the frame-building organisms do not necessarily or even characteristically comprise the bulk of the reef. They merely serve to hold it together, and the frame they build is a trap for clastic or chemically precipitated sediments. Sollas (1904, p. 6) aptly refers to the Funafuti reef as "a coarse 'sponge' of coral with wide interstices, which may be either empty or filled with sand"—a description that is widely appropriate for other existing reefs. Important organic components of reefs, according to locality and time, are corals, calcareous algae, stromatoporoids, sponges, bryozoans, the aggregated tubes of sessile gastropods, calcareous worm tubes, foraminifera, and, of course, the whole gamut of organic detritus.

The part played by calcareous algae in the formation of existing and some ancient reefs is conspicuous to dominant—crustose and articulate corallines, as well as various green and blue-green algae contributing to the reef mass through secretion, precipitation, or sediment binding. Umbgrove (1947, p. 731) notes existing reefs near Timor and Tertiary reefs in Java that were built primarily by nodular coralline algae; Mme. Weber van Bosse (Weber van Bosse and Foslie, 1904, pp. 4–7) describes and illustrates extensive reefs and banks along the east side of Borneo and in the Sulu Archipelago that appear to be formed almost exclusively of nodular and branching crustose corallines; patch reefs formed exclusively of branching and nodular corallines occur at Saipan in the Mariana Islands; and algae were at least locally important Paleozoic and pre-Cambrian reef- and bioherm-builders (Adams and Frenzel, 1950, pp. 302, 305; Cloud, 1942; Cloud and Barnes, 1948, pls. 18–19; Hadding, 1941, 1950; Johnson, 1942).

On the other hand, the role of algae in existing reef construction has been exaggerated for some localities. An example is Onotoa Atoll, in the southern Gilbert Islands, where Setchell (1928, p. 1840) and Finckh (1904, p. 147) report encrusting coralline algae as the principal builder of the seaward reef. While it is true that algae do extensively veneer the outer reef at this atoll and are the principal growth form of the windward part of the reef in its present essentially static condition, the frame and undermass of the reef, wherever the surface veneer of algae is broken away, are seen to be principally of corals and mainly the genus *Heliopora* (Cloud, 1952, p. 22, 63, Fig. 3, profile 5). Such occurrences as that at Onotoa led Finckh (1904) to the erroneous belief that "corals are not active reef builders" and persuaded Setchell (1928, pp. 1841, 1843) to the more defensible though debatable conviction that "the nullipore [crustose coralline algae] is *the* controlling organism in reef formation in the central Pacific."[3]

At certain times and places sponges have been important frame-builders in reef construction, and the distinctive supposed cnidarians known as stromatoporoids (Lecompte, 1951, pp. 36–39) were major contributors to the frames of Paleozoic reefs. The stromatoporoids in many ways played a part similar to that of the crustose coralline algae ("*Lithothamnion*") in existing reefs, and notably abetted the rather weakly attached Paleozoic corals in building strong reef structures (Hadding, 1941, is a random illustration from numerous possible references).

Other cnidarians, besides corals and stromatoporoids, that play or have played an important part in reef growth are hydrozoans. *Millepora* is and was an active and bulky contributor to the frames of reefs at Funafuti (Finckh, 1904, pp. 129, 125) and elsewhere (Gardiner, 1931, pp. 89, 171; Agassiz, 1903, p. 267). And among corals themselves the "typical" reef-building hexacorals of most reports, the scleractinians, are out-bulked as near-surface frame-builders in several and perhaps most of the Gilbert and Ellice islands by the blue alcyonarian *Heliopora* (Cloud, 1952; David, 1904, pp. 45–46, 51; David and Sweet, 1904, pp. 66, 69–70; Finckh, 1904, pp. 135–36). Also the more loosely spiculate alcyonarians, which normally contribute only their distinctive calcareous spicules to the interstitial sediments, at places participate in frame-building by the formation

[3] Bracketed insertion and italics are added to Setchell's statement.—Preston E. Cloud, Jr.

of large basal masses of solidly matted spicules, for example, *Sclerophytum* at Samoa (Cary, 1931, p. 97).

The intertwining sedentary gastropod *Vermetus* collaborates with encrusting coralline algae, and rare corals, to form thin bench-veneers that simulate organic reefs in the Cape Verde Islands (Crossland, 1905), and similar associations are not uncommon in the warmer seas. The calcareous tubes of gastropods, pelecypods, and annelids at many places contribute incidentally to reef formation, and locally may add significantly to the reef mass.

Bryozoa, also, at places add to the materials of existing reefs and were important contributors to some Paleozoic reefs.

Even siliceous annelid tubes have formed extensive and truly wave-resisting fringing and barrier reefs in the North Sea and across the bay of Mt. St. Michel since 1914 (Richter, 1927)—structures which are probably in part analogous to some of the *Scolithus* beds of Paleozoic time.

Growth of organic reefs may proceed at a relatively rapid rate—easily enough to keep ahead of even rather heavy accumulation of débris behind them. Vaughan's classic studies on rate of coral growth led him to the conclusion that "reef corals can build a reef 46 meters (150 feet) thick within a period ranging from 1,800 years to 7,500 years; but, in places, a reef of such a thickness might be formed within 1,000 years" (Vaughan, 1919, p. 215).

Those desiring to make direct comparisons with data from drilling in reefs and reef environments should consult the reports of Hinde (1904), Kuenen (1947), Ladd and associates (1948), and references there cited.

Limiting Ecologic Factors

The probable ecology of ancient organic reefs is indicated by various associated organisms, as well as by physical criteria. Among organic factors the corals and algae have the broadest and most general significance and are therefore discussed briefly. Foraminifera appear to be of limited general application in reef studies, and for specific problems within the range of known ecologic patterns the current techniques are so well known that elaboration would be trite. Other possible criteria for reef paleoecology are noted where appropriate, without special discussion.

REEF-BUILDING CORALS

For more than a century it has been recognized that reefs to which stony hexacorals (Scleractinia) contribute importantly are in themselves presumptive evidence of growth in shallow, tropical, or subtropical waters (Quoy and Gaimard, 1824, p. 661; Couthouy, 1844, p. 76; Darwin, 1889, pp. 108–18; Dana, 1872, p. 118). For active reef building the depth should not exceed 25 fathoms; for really vigorous growth 15 fathoms is the normal limit; and the temperature must not fall below 18.5° C. or remain long even at that figure (Vaughan, 1919, pp. 197, 202; Vaughan and Wells, 1943, pp. 52, 55). Very recent specific data are for

Bikini, in the Marshall Islands, where modern reef-building corals are most varied and abundant above 15 fathoms, diminish rapidly to about 45 fathoms, and fall off to only one or two species at 85 fathoms (Ladd *et al.*, 1950, p. 424).

Even more impressive than conclusions reached from physical measurements of light penetration, temperature, and numbers of species found at different depths are the results of direct observation. It is a common experience in diving along the fronts of reefs to observe an abrupt diminution in variety and abundance of corals below 5–10 fathoms, depending on local conditions. In the relatively turbid waters of the Bay of Batavia, Verwey determined the actual compensation point of oxygen production from algae suspended in jars, and measured the lower limit of the living reefs by letting a float rise to the surface from the base of living reefs. He found the depth of the oxygen compensation point to range from 6 to 11 fathoms and the base of the living reef to lie at only 4–8 fathoms (Verwey, 1931a, pp. 194–95). At least locally in the Bahama Islands reef building terminates at similar depths (Newell *et al.*, 1951, p. 23). In the lagoon of the Great Barrier Reef the base of active coral growth is around 6–7 fathoms (Fairbridge, 1950, pp. 334, 363). And in the open Pacific Ocean 8–10 fathoms appears to be a common dimunition point.

Light and temperature are the controlling factors—light through its effects on the photosynthetic activities of associated algae (commensal zooxanthellae, and frame-building corallines), and both light and temperature on the phototropic, stenothermal, reef-building corals themselves. The importance of the general limitations thus placed on the formation of existing reefs is not minimized by such facts as the local occurrence of small "korallenbänke" at depths as great as 100 meters in colder seas, or of thriving colonies of non-reef-building *Lophohelia* at 200–600 meters depth in the Trondhjem Fiord (Pia, 1933, p. 255). However, the geologist concerned with the interpretation of ancient organic colonies must keep such facts ever in mind, in order to avoid drawing conclusions based on false analogy.

With few reservations Vaughan's 1919 conclusions on the conditions necessary for the vigorous growth of reef-forming corals are still acceptable. The conclusions are: "(a) Depth of water, maximum, about 46 meters (25 fathoms); (b) bottom firm or rocky, without silty deposits; (c) water circulating, at times strongly agitated; (d) an abundant supply of small animal plankton; (e) strong light; (f) temperature, annual minimum not below 18° C.; minimum average temperature for the coldest month in the year not lower than about 22°C. (72°F.); (g) salinity between about 27 and about 38 parts per thousand" (Vaughan, 1919, p. 215).

The reservations concern points a, b, and c:

(a) Whereas vigorous reef growth does not ordinarily extend below 25 fathoms and is commonly limited to shallower depths, it seems probable that reefs may grow upward from depths as great as 45–50 fathoms under specially favorable conditions. Thus, although reefs more than 150 feet thick probably underwent subsidence (or rise of sea) during growth, under special conditions reefs might attain thicknesses as great as 300 feet without subsidence. On the other hand, depths of only 60–75 feet seem to be more usual maximums for vigorous reef building. Where the matter is in doubt it might in some instances be resolved by careful study of the vertical succession of types of reef-building organisms.

(b) Specially adapted types of corals (such as species of *Porites*, *Acropora*, and *Pocillopora* among living and Pleistocene genera) may begin growth on bits of shell or pumice pebbles on silty or even muddy bottoms and from there spread out to build the firm base from which a reef may be erected—though it is, of course, true that fine sediment in transportation is inhibitive to coral growth. The coral reefs of the Java Sea and the Bay of Batavia are rooted in a muddy bottom and surrounded by relatively turbid waters (Umbgrove, 1949, p. 7; Verwey, 1931b, p. 200). They are believed by Umbgrove to have developed in the manner suggested, although the possibility of their being rooted on a more solid floor beneath the mud appears not to be conclusively eliminated.

(c) With regard to circulation of water, the prime factor is frequent enough change to bring nutrients, maintain the oxygen supply, and remove waste and fine sediments. Agitation is not required, although circulation competent to do the job specified is ordinarily accompanied by surface agitation.

By analogy the foregoing data and conclusions may be applied to interpretation of Mesozoic and Tertiary reefs in which corals functioned as frame-builders. These corals were of the same general sorts as existing corals and were very likely subject to similar controls. Depths inferred may be modified according to evidence of probable degree or absence of turbidity.

The evidence that corals provide about environment of Paleozoic reef formation is mainly indirect. Large masses of organically formed calcium carbonate, regardless of age, imply reasonably warm and shallow waters, because of the high calcium metabolism indicated. And Paleozoic corals and associated organisms did form structures analogous to existing reefs. Firm affirmation of the probable limiting depths and temperatures of Paleozoic reef formation is, of course, not warranted without evidence other than their mere existence. On the other hand, such evidence is easier to adduce and more reliable than may be commonly supposed, and the algae are of special significance in this regard.

CALCAREOUS ALGAE

Calcareous red and green algae are common not only in existing reef environments; they also contributed significantly to fossil reefs, ranging well back into the Paleozoic. The calcareous red algae are the corallines, including crustose forms which range back to the Lower Cretaceous and articulate forms which range at least as low as Upper Cretaceous (Pia, 1927, p. 100–04). The extinct Solenoporaceae, ranging from Lower Ordovician to Jurassic, are presumably the Paleozoic precursors of the modern crustose coralline algae. The calcareous green algae include the living family Dasycladaceae, which ranges back to Early Ordovician with little variation in general structure, and the Codiaceae, which ranges certainly from Permian and possibly Ordovician to Recent (Johnson, in Smith *et al.*, 1951, p. 194). The Charophyta, also calcareous green algae with a long fossil record, are mainly non-marine. The blue-green algae appear to be represented as fossils by laminated and occasionally filamentous, but mostly structureless crusts that locally formed or contributed to organic reefs and bioherms of the past. The distribution of algae in the post-Silurian Paleozoic rocks is summarized and tabulated by Pia (1937) and by Johnson (1946).

Algae are of significance in paleoecologic interpretation because certain groups are limited to particular latitudes and temperatures, and all are restricted by their need for light to perform the vital function of photosynthesis. The depth of water in which algae may live thus varies not only according to clarity of water, but also according to intensity and obliquity of sunlight (oblique light is more readily refracted and must travel farther to reach the same depth), and probably according to the wave lengths of light used in photosynthesis.

The ordinary limit for vigorous photosynthetic activity has already been given as 25 fathoms, and the general "floor" for marine algae is approximately 50 fathoms (David *et al.*, 1904, pp. 153, 158; Lemoine, 1940, pp. 83–84, 116–17, 127–28; Feldman in Smith *et al.*, 1951, p. 314). This is a valid and applicable generalization in terms of present knowledge, but interpretive refinement can be made according to the groups of algae studied. Red algae survive over a much greater range of latitude and depths in present seas than do green algae, and relationships of both reds and greens with fossils ranging back to Early Ordovician are close enough to make reasonable the inference of approximately similar restrictions in the past. The critical information is mostly given in an excellent summary by Mme. Paul Lemoine (1940).

The crustose coralline algae, which have the greatest latitudinal and depth range of existing calcareous forms, range from the tropics into temperate and cold seas, with five of the ten most generally recognized existing genera restricted to the tropics and five ranging from tropic to cold waters. As regards depth, of 59 general records accepted as authentic by Lemoine (1940, pp. 83–84), more than half (34) were from less than 60 meters, all but 7 were from less than 100 meters, and most below 60 meters were tropic records. The deepest record accepted by Lemoine as authentic for the calcareous algae was 156 meters (507 feet), from Funafuti, even though David and associates (1904, p. 156) state that at Funafuti a thinly "encrusting type of *Lithothamnion* . . . lives from low-water spring tides down to at least 200 fathoms." Gilbert Smith (1938, p. 296) also doubts the reliability of records of living algae below 200 meters, and Jane Feldman (in Smith *et al.*, 1951, p. 314) concurs with Smith and gives her deepest record for living algae as 180 meters in the clear waters off the Balearic Islands. Dredging on the outer slopes of Bikini Atoll (Ladd *et al.*, 1950, p. 423) yielded algae from "below 500 feet." After considering all records, as well as physical factors (Cloud, 1942,

p. 370), it seems that a depth of 200 meters is the extreme limit at place of growth indicated by algae of any sort, but that much shallower depths are normal and that branching or nodular corallines are rarely abundant below 30 meters. It further appears that red algae may be able to grow to greater depths than green algae because the red pigment (phycoerythrin) enables them to use blue light in photosynthesis at depths below which the red and yellow rays are screened out (David *et al.*, 1904, p. 156; Smith, 1938, pp. 297–98; Strain, in Smith *et al.*, 1951, pp. 258–60).

Among the articulate coralline algae, known depth range is much less than for crustose types. Of 39 general records accepted by Lemoine (1940, pp. 116–17), all but 8 are from less than 30 meters and the maximum depth recorded is 90 meters. The articulates, like the crustose corallines, range from the tropics to the polar regions (71° N.), with certain genera typical of the tropics, and others ranging widely.

In contrast to the coralline algae the calcareous greens are primarily tropical and subtropical, with a maximum range to 46° N. and 34° S. for the existing and Tertiary genus *Halimeda* (Barton, 1901, p. 7; Pia, 1937, Fig. 1; Lemoine, 1940, p. 119). Members of the family Dsaycladaceae are very abundant between 3 and 5 meters depth in tropical and subtropical seas and rarely range below 30 meters. The extreme depth recorded for a dasycladacean is 70 meters (Lemoine, 1940, pp. 122, 124). The family Codiaceae, to which *Halimeda* belongs, appears to be most distinctive of waters shallower than 95 meters, but has maximum recorded depth of 120 meters (Lemoine, 1940, pp. 119–20).

To summarize in fathoms, algae as abundant contributors to reef formation virtually establish construction at depths of considerably less than 110 fathoms—they strongly indicate depths of less than 50 fathoms, very probably less than 25 fathoms, and ordinarily less than 15 fathoms (much less depths are indicated for fresh water). More specific lower limits for depth of growth are indicated as follows: crustose corallines (Lower Ordovician?, Cretaceous-Recent), 110 fathoms maximum, 15 fathoms suggested by abundant branching and nodular forms; articulate corallines (Upper Cretaceous-Recent), 46 fathoms maximum, 14 fathoms probable; dasycladacean greens (Lower Ordovician-Recent), 38 fathoms maximum, 14 fathoms probable, 2–3 fathoms likely; and codiacean greens (Ordovician?, Permian-Recent), 65 fathoms maximum, 45 fathoms probable. Calcareous blue-greens, the primary builders of fresh-water reefs and presumably of the pre-Cambrian to early Paleozoic bioherms, probably do not form large structures below about 10 meters in fresh water and 30 meters in the sea (Cloud, 1942, p. 371).

As regards temperature, the coralline algae have significance only in terms of suggestion of warm-water formation (they are locally abundant in Norwegian fiords) where they comprise large masses of calcium carbonate, or by the occurrence of particular genera in Cenozoic deposits. Tropical to subtropical temperatures, however, are indicated by the occurrence of calcareous green algae.

Facies Relationships of Principal Reef Types

SUBLINEAR REEFS

Fringing reefs, barrier reefs, and atolls are and probably were basically wave-breaking and surf-fronting structures. That they grew upward to the surface wholly within the zone of wave action is strongly indicated by their approximately linear form, especially for atolls and barrier reefs. Barrier reefs and the outer reefs of atolls are likely to be very irregular, grading to discontinuous lines and areas of patch reefs, where they form in protected situations. Fringing reefs, on the other hand, are likely to develop as well along protected shores because they mantle the shore line—which is, or course, in the zone of whatever wave action there is. Factors controlling development of faros (Maldives) as well as anastomosing to loosely reticulate or relatively extensive linear lagoon reefs (Canton Atoll) or bank reefs (Johnston Island) are not well understood. However, it may be significant that the best known examples are located within large atolls or on open banks that are freely exposed to wave action. Faros might develop from patch reefs by more vigorous growth of outer margins, whereas development of linear lagoon or bank reefs may well be importantly influenced by current and wave action over broad lagoons or banks. Such linear reefs could also represent structure that developed on open banks before the final upgrowth of the outer atoll ring-reef, or perhaps successive organic ridges built surfward across a bank.

Reef structures of sublinear plan in the present oceans presumably have complicated lateral relationships at depth. Most have probably grown upward from Pleistocene sea-level, with interruptions of growth, and with irregularities induced by local tectonic movement and planation at various stages. Such reefs are and presumably were surf-fronting and may thus be expected to include many broken and displaced organic structures, much interstitial clastic débris, varying proportions of chemically precipitated calcium carbonate (loss of CO_2 from sea water because of agitation, photosynthesis, and evaporation), and not necessarily a dominance or even marked abundance of organic structures in position of growth. As reef deposits to which scleractinian corals or algae contribute importantly would not ordinarily accumulate to thicknesses much exceeding 150 feet without subsidence, such thicknesses are presumptive evidence of subsidence. Morphology and gradational relationships of the linear reefs vary according to type and location of reef, history with reference to sea-level, slope and nature of adjacent or near-by coasts, prevailing wind conditions, depth of offshore troughs or basins, and source and nature of sediments supplied to offshore troughs or basins. The possible variations are virtually infinite, and only a few greatly generalized and simplified forms are discussed.

FRINGING REEFS

Fringing reefs (Fig. 1) are the expectable reef forms of shores that are stationary (Fig. 1C), slowly rising (Fig. 1D), or rapidly subsiding (Fig. 1B) with

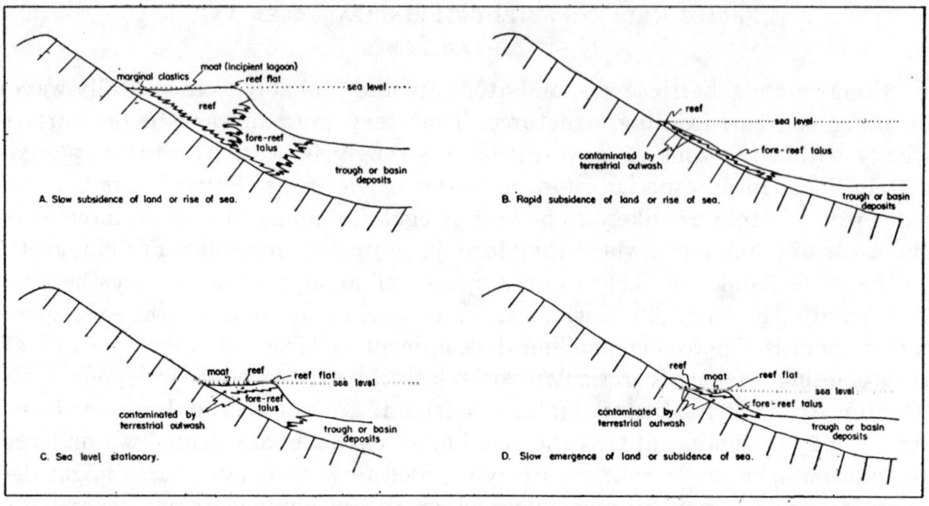

FIG. 1.—Conjectural relationships of fringing reefs under conditions of subsidence, stability, or emergence of land relative to sea.

respect to sea-level. Slow subsidence (either gradual or episodic) may produce a fringing reef similar to that in Figure 1A, but, as Darwin long ago pointed out, if subsidence continues from that stage a barrier reef should result. Fringing reefs may be expected to show the form of irregular calcareous mantles, veneers, or wedges with abrupt contacts against the sloping substratum on which they grew —according to Vaughan (1919, pp. 229, 237), basal unconformities are the rule for fringing reefs. Organic structures in position of growth are to be expected, but may not be abundant.

In the offshore direction, fringing reefs should interfinger with or overlap generally steep, sea-facing débris slopes composed mainly of broken and rolled fragments of the reef material mixed with débris from the adjacent shore. Eventual burial of such submarine talus slopes is likely to produce sharp contacts, if the slopes are in an environment where abundant terrestrially derived débris is available for their burial. In an essentially open sea environment, however, gradation from reef talus to bedded limestones may be over a wide interval.

On exposed windward shores of high islands, fringing reefs may grow somewhat above high tide to splash-level of the surf, developing a special surface morphology of mutually impinging rimmed basins and benches reminiscent of hot spring or geyser deposits. Over these the splash descends in constant flow and agitation, providing ideal conditions for precipitation of calcium carbonate through escape of carbon dioxide and the abundant growth of algae, tube-building annelids, and other encrusting organisms. Such *terraced ramps* and adjacent *terraced pedestals* (probably around cores of differing nature), if preserved by burial, would not only show abrupt contacts with adjacent deposits but would also suggest their site of formation to have been an exposed windward shore.

The fronts of many fringing reefs are separated from the beach by very shallow and narrow bodies of water, long known as *moats*. Their bottoms are apt to be uneven, with small irregular mounds of coral growth or rock projecting toward or above the surface, and a veneer of detrital calcareous fragments. Elevated moats and moat-deposits are rather obvious features of some elevated fringing reefs of probably late Pleistocene or Recent age in the western Pacific (Saipan, Tinian), but there is serious doubt whether really ancient examples could be recognized as such.

BARRIER REEFS

The barrier reef (Fig. 2) lies offshore from the land, being separated from it by a lagoon of varying depth and area. It is here regarded as the characteristic land-encircling reef form initiated by submergence (Fig. 2a), but it may also grow upward from a pre-existing platform within growing reach of sea-level or even under conditions of emergence (Hoffmeister, 1932, p. 38). A pre-existing platform of organic growth (Fig. 1A or 1C), inorganic construction, or abrasion is held necessary for initiation of the barrier-reef form. This platform is, of course, likely to be unmarked or obscure where structurally organic or bioclastic in origin. With proper prior conditions, barrier reefs may persist under conditions of stationary sea-level (Fig. 2B) or emerging land (Fig. 2C). Some barrier reefs grade laterally to fringing reefs, while their lagoons concurrently change to moats (west coast of Saipan).

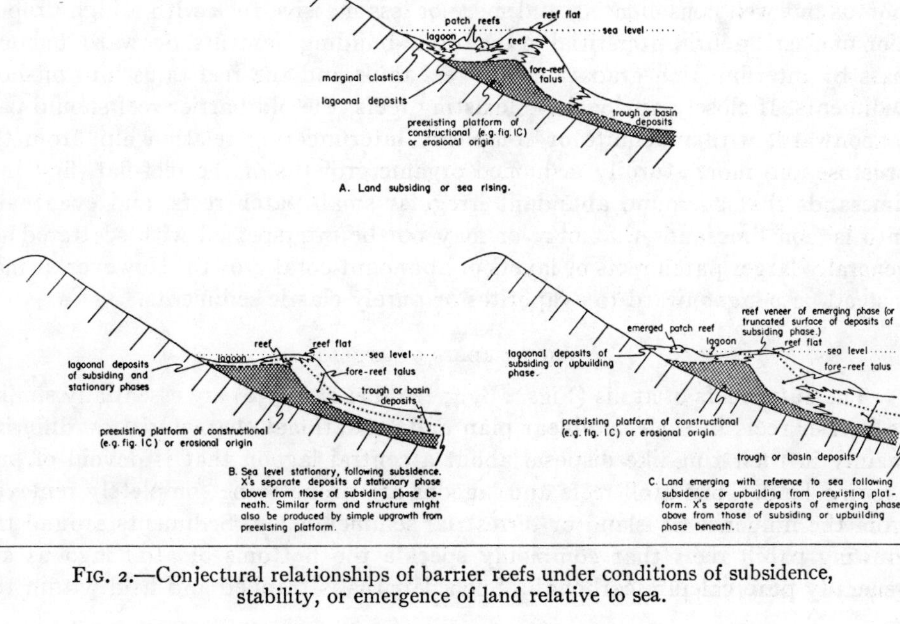

FIG. 2.—Conjectural relationships of barrier reefs under conditions of subsidence, stability, or emergence of land relative to sea.

Barrier reefs are linear, wave-fronting structures that are essentially similar to fringing reefs. Continued slow submergence of a fringing reef, or episodic submergence in steps not ordinarily exceeding 15–25 fathoms, may lead to development of a barrier reef, and it is difficult to be sure in some instances whether the structure is properly a fringing reef or a barrier reef. Barrier reefs are built of the same sorts of organisms and clastic or physico-chemically precipitated calcium carbonate as are fringing reefs, and they show a similar range in proportions. Their offshore characteristics are similar to the characteristics of fringing reefs except that they are farther removed from the contamination effects of land, and although burial may be by means of derived terrestrial detritus, the reef rock itself should be relatively pure calcium carbonate.

Gradation from barrier reef into lagoonal sediments may range from sharp to indefinite. The form and structure of the reef-flat is generally rather sharply distinct from the lagoonal sediments and back-reef growth in plan view of present reefs, but mappable gradations of ancient reef environments, especially where the third dimension is viewed, may become complex.

Behind the reef-flat in existing seas there may be several zones of decreasing concentrations of corals of different sorts that grade complexly into the lagoon proper, the latter being characterized by a floor of clastic limesand surrounding occasional to abundantly scattered patch reefs of varied sorts as well as fields of irregularly growing corals and algae. Areal fluctuations of the reef-flat with time, related to submergence, emergence, and growth factors, should produce contacts with lagoonal sediments that are indefinite or interfingering. In three dimensions the ancient barrier reef itself should be an irregular ridge, wedge, or mantle of porous but well consolidated, and more or less massive rock with a high proportion of clastic débris interstitial in the frame-building elements. Seaward it should pass by interfingering gradation into reef talus and the reef talus into offshore sediments. If closely analogous to existing reefs, the old barrier reef should pass lagoonward, with indefinite or complexly interfingering relationship, from the crustose and more sturdily fashioned organic growths of the reef-flat, first into limesands that surround abundant irregular small patch reefs, and eventually into lagoon limesands that may or may not be interspersed with scattered but generally larger patch reefs or layers of abundant coral growth. However, it may instead pass lagoonward to evaporites or purely clastic sediments.

OUTER REEFS OF ATOLLS

The outer reefs of atolls (Figs. 2B, 4; Ladd *et al.*, 1950) are essentially similar to barrier reefs in their sublinear plan and gradational characteristics, differing mainly in their ring-like disposal about a central lagoon that is devoid of preexisting land. Thus atoll reefs and lagoons are prone to be completely removed from the influence of island or terrestrial sedimentation. Sediments around the growing patch reefs that commonly speckle the bottoms of atoll lagoons are generally pure calcium carbonate. Such detritus is trapped and held within the

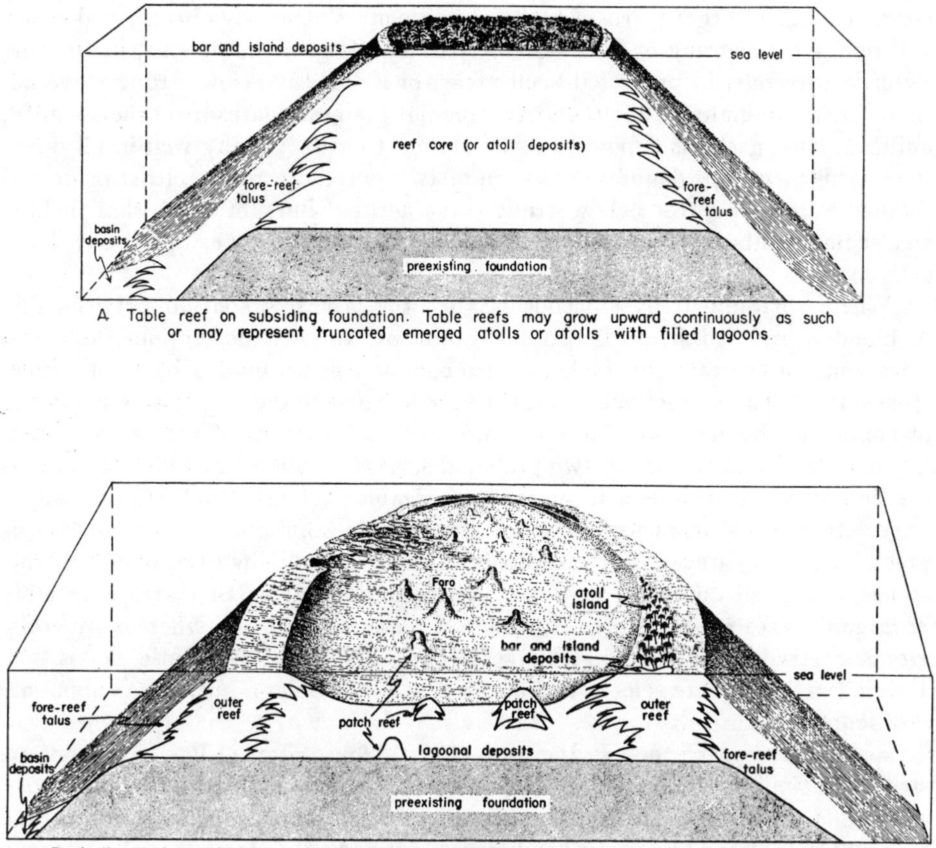

A. Table reef on subsiding foundation. Table reefs may grow upward continuously as such or may represent truncated emerged atolls or atolls with filled lagoons.

B. Atoll with lagoonal patch reefs on subsiding foundation (or, in post-Paleozoic, on one initially not deeper than about 25 fathoms.)

FIG. 3.—Conjectural relationships of atolls and table reefs.

outer ring reef, which Sollas (1904, p. 27) has described as similar to a retaining wall around an area of loose deposits.

Unless an atoll has grown upward from a pre-existing platform, it can not form except by submergence. Judged from the thicknesses of shallow-water sediments beneath atoll reefs that have been penetrated by the drill (summarized by Ladd *et al.*, 1948; Shepard, 1948a), and from the general occurrence of atolls in the subsiding parts of the Pacific Ocean (Darwin, 1889, frontispiece; Shepard, 1948a), most existing atolls have formed under conditions of subsidence.

PATCH REEFS AND SIMILAR STRUCTURES

In the geologic column the generally small and subequidimensional to irregular organic mounds, knolls, pinnacles, pedestals, or platforms known as patch reefs (Figs. 2, 3) or bioherms are ordinarily of simple shape and structure. Their lateral change to other types of rocks or sediment is generally abrupt, and either narrowly interfingering or even. Living patch reefs grow in lagoons, in the surf lee of linear reefs, in protected shelf areas, or at depths below strong wave action. Their dominantly subcircular to irregular plan originates from their ability, unlike organic growths exposed to surf action, to grow equally well in all directions. This form thus denotes growth in relatively quiet waters—either protected shallow waters or water below strong wave action—but, for those that include algal components or reef-building hexacorals, within the zone of photosynthetic activity.

Where surrounded by siliceous clastics the lateral gradation between reef or bioherm and sediments is ordinarily abrupt. In geologic sections, however, calcareous patch reefs and bioherms are commonly surrounded by clastic limestones. In such event distinction must be made between the wholly clastic texture of commonly bedded surrounding sediments and textures of the reefs or reef-like masses. The latter are of two principal sorts: (1) structures which are in part clastic but which include a framework of organic skeletal structures in position of growth, and (2) masses of very fine-grained and laminated calcareous organic precipitates or aggregates. It is difficult to make such distinctions between limestones from well cuttings. However, where individual cuttings are invariably or largely pieces of single organisms a reef mass is suggested, whereas generally granular structure of cuttings suggests entirely or largely clastic deposition. Dolomitization, where selective, may resolve the problem, but more commonly it obscures it.

Some of the patch reefs in the protected shallow waters of Pacific lagoons are strikingly similar to the Middle Devonian patch reefs of Michigan (Lecompte, 1938, pp. 11–17, Pls. 2–3; Dunbar, 1949, Fig. 129) in shape, general structure, and observable relations to adjacent sediments. A principal difference is that stromatoporoids instead of crustose algae are the primary frame-building consorts of the Devonian corals. The Devonian patch reefs alluded to may not have grown in lagoons, but their shape alone is strong evidence of growth in relatively quiet waters, and their composition is indicative of relatively warm and shallow water. In present as in ancient sediments, certain types of coral and other organic growth hold together better than other sorts, with consequently variant degrees of sharpness in lateral gradation away from the reef or biohermal structures. The Michigan reefs noted show such variations. Patch reefs that are conspicuously flat-topped, or, especially, flaring-topped, may be considered indicative of wave-breaking structures. Some of the Devonian patch reefs in the quarries at Alpena, Michigan, have flaring upper margins.

In the ancient epeiric seas of North America, patch reefs appear to have been the dominant reef type; and their limits are for the most part readily delineated, at least so far as surface mapping is concerned. Similarly shaped masses of clastic structure, however, or masses that appear similar in cross section or to the drill, occasionally result from the piling up of locally derived bioclastic débris where such material is surrounded by bottom of a different nature and fill on surrounding bottom is so balanced against upgrowth of the bioclastic mound as to prevent conspicuous intermingling of the two. Such may be the origin, for instance, of certain mounds, lenses, or irregular masses of crinoidal débris within successions of shale (Cloud and Barnes, 1948, Fig. 3) or different types of limestone.

TABLE REEFS

Table reefs that are not the result of lagoon-filling or truncation of atolls may be considered the patch reefs of the open ocean (Figs. 3, 4). Such table reefs differ from ordinary patch reefs primarily in larger size and their oceanic environment. The characteristics of their gradation to adjacent sediments should differ from those of many patch reefs in displaying an outer zone of fore-reef talus (due to their larger size and to their exposure in later stages to strong wave action). Table reefs that do not represent filled or truncated atolls presumably grew upward from depth, for growth wholly within the zone of oceanic wave action should produce linear structures.

That some, and perhaps most, table reefs, however, do represent modified former atolls is strongly suggested by comparison between charts of table reefs in various stages of development and atolls with shoal lagoons. Figure 4 illustrates some examples, and the National Geographic Society's map of the Pacific Ocean (September, 1943) suggests analogies between lagoonless Enderbury, Vostok, Starbuck, and Flint islands and the atolls of Gardner, Sydney, and Malden. Several Gilbert Island atolls, such as Onotoa, Tabiteuea, and Beru, require only a little shoaling and truncation, or a little more lagoon-filling, to become table reefs, and the islands of Tamana and Arorae, already table reefs, may well be so merely because they were smaller than ordinary atolls to begin with (*United States Hydrographic Office Chart 0119*). Judged from the description given by J. P. Couthouy (1844, p. 141) as well as from maps and charts, Christmas Island, in the northern Line Islands, may exemplify both a nearly filled atoll and a very large incipient table reef (National Geographic Society map; Hydrographic Office charts). Gardner, Washington, and Enderbury islands, all shown at the same scale in Figure 4, provide an interesting sequence from typical atoll, to table reef with a relict "lagoon," to table reef with merely wet internal depressions.[4] Swains and Sydney islands are atolls with closed central lagoons, and the water of the lagoon at Swains is fresh, at least at certain times in the surface layer (L. P. Schultz, 1943).[5] Birnie, McKean, Niutao, and Phoenix islands are table reefs with a central depression apparently generally similar to that of Enderbury. Baker, Howland, and Nurakita (Fig. 2) are table reefs in the strictest sense.

[4] Leonard P. Schultz, who visited Enderbury Island during May, 1939, informs the writer that the large depression south of the island's center was then floored with sticky calcareous mud and brackish water not exceeding knee depth.

[5] In oral communication Schultz states his belief this was due to heavy rains and that the lower layers remain saline.

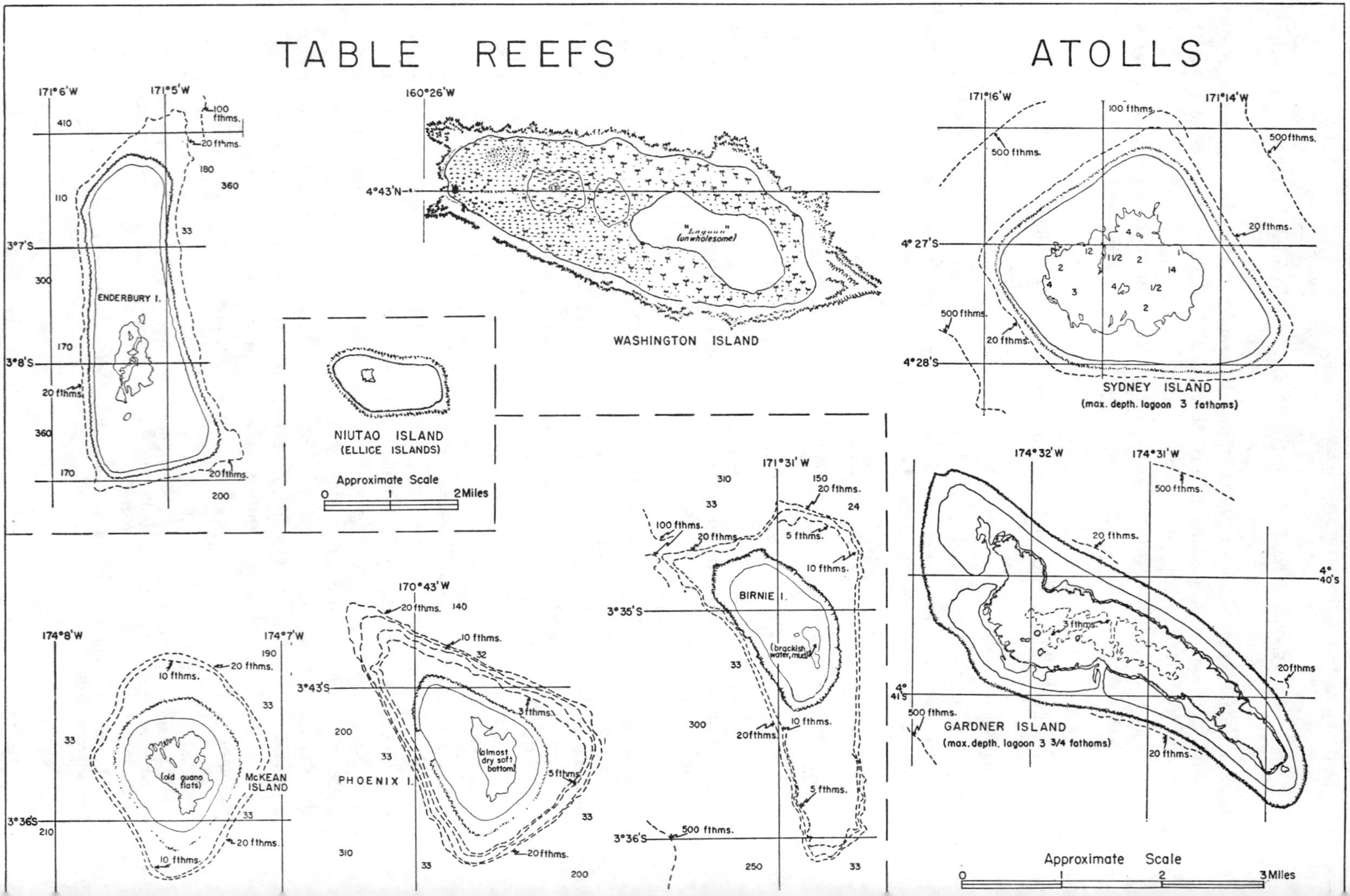
TABLE REEFS
ATOLLS
ENDERBURY I.
WASHINGTON ISLAND
"Lagoon" (unwholesome)
NIUTAO ISLAND
(ELLICE ISLANDS)
Approximate Scale
0 1 2 Miles
SYDNEY ISLAND
(max. depth. lagoon 3 fathoms)
BIRNIE I.
(brackish water, mud)
GARDNER ISLAND
(max. depth. lagoon 3 3/4 fathoms)
McKEAN ISLAND
(old guano flats)
PHOENIX I.
(almost dry soft bottom)
Approximate Scale
0 1 2 3 Miles

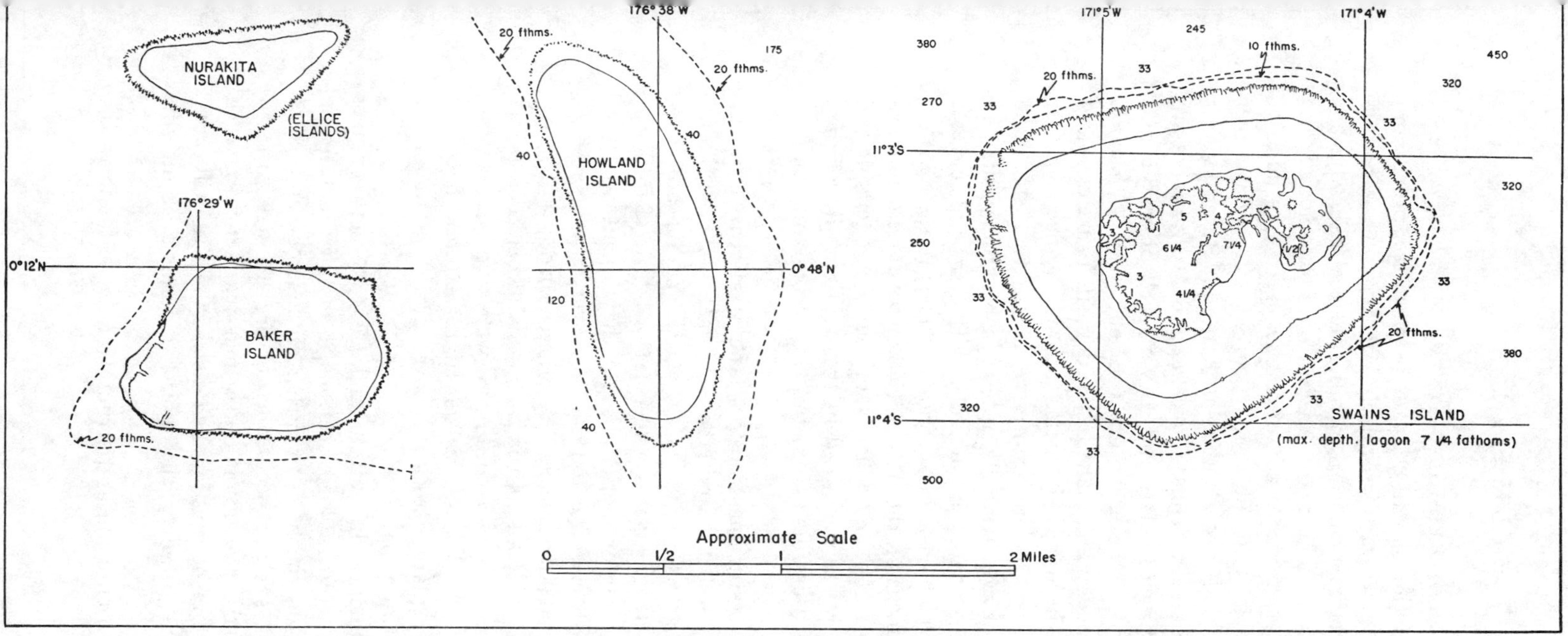

FIG. 4.—Comparative charts of table reefs and atolls with shoal lagoons. Data from various published charts of Hydrographic Office, United States Navy. All soundings in fathoms.

Recognition of Reef Structure

In studies of linear reefs, distinction should be made between reef proper and reef complex. The reef-flat and its downward irregularly interfingering or veneering extension is the reef proper. The reef complex includes the reef-flat, adjacent patch-reef-studded parts of the lagoon floor, and the seaward talus slope.

Distinctive species or specific assemblages are, of course, useful in studies of reef facies and environment, especially in the post-Paleozoic. In Cenozoic reef deposits, for instance, the sort and degree of wear of Foraminifera and other organisms are indicative of environment. Thus, *Calcarina* and *Marginopora* are characteristic inhabitants of existing reef-flats and, if very well preserved and abundant, are presumptive evidence of a reef-flat environment. Worn *Calcarina*, however, are at many places extremely abundant in the deposits of beaches and atoll islands and locally common in lagoon deposits. Conversely, deeper-water benthonic Foraminifera may be used as presumptive evidence to exclude the reef-flat environment, although they occasionally appear in beach deposits. An example is *Cycloclypeus*, which ranges through much of the Tertiary and suggests a depth of about 100 meters, but which has a known range from 37 to 1,420 meters and has been found as dead specimens on the beach at Okinawa (oral communication from F. S. MacNeil). Foraminiferal evidence is, of course, less reliable and more difficult to evaluate as one works backward in the stratigraphic column, although Henson (1950) has found it useful in studies of Cretaceous and Cenozoic reefs in the Middle East.

On the other hand, the relative sturdiness of coral and algal growth of any geologic horizon should serve fairly well as a limited environmental index, abundance of either delicately branching growths or very large massive coral heads being indicative of quiet waters and evidence against a reef-flat or high reef-front environment. Strong, stubbily branching or encrusting growths, in contrast, characterize the reef-flat and upper reef-front. Growth form, moreover, varies markedly with environment even within the same species. Vaughan (1919, pp. 195–96, Pls. 5, 7, 8) describes and illustrates striking variations in habit of three species and genera of common existing reef corals according to wave exposure. In one of these remarkable examples, a piece of coral broken from a stumpy flattened boss on the exposed reef and transplanted to a lagoon environment thereafter assumed a delicately branching growth pattern.

A differentiation that is commonly difficult to make is that between reef proper and reef talus. In well exposed geologic sections it should generally be possible to make this distinction on the basis of rather obvious characteristics of rock structure. Gross features, such as occurrence of organic structures in position of growth or the contrast between coherently massive reef framework and chaotic talus structure, should be helpful for such distinctions. In drill cores, cuttings, or poorly exposed sections, however, the differentiation is more difficult and may be impossible.

A promising key to primary orientation and structure may be found in the use of composite and incomplete internal fillings. The upper plane surfaces of partial fillings of organic or non-organic cavities in coral growth or elsewhere, or the planar contacts in composite fillings where clear secondary calcite has come to occupy the space above the primary partial filling, denote horizontality at the time of primary partial filling (Cullison, 1938) or at the time of coming to rest of a block that contained loose fine débris in such cavities. Thus, non-horizontal orientation of such surfaces in cores or in outcrop would indicate displacement subsequent to partial filling and induration of filling material, and, where common, would be evidence of location in the talus zone. Likewise, orientation of surfaces of incomplete or composite fillings of such openings found to be at variance with primary horizontality indicated by growth pattern would suggest displacement of the block in question while the filling material was still loose and free to assume parallelism with a new plan of horizontality at the place of coming to rest. This also, where common, or in combination with non-horizontal orientation of similar or other incomplete or composite fillings, would suggest reef talus. On the other hand, predominantly subhorizontal arrangement of the surfaces of primary partial fillings of cavities in frame-building organisms, in combination with their common orientation in normal position of growth, would be fairly conclusive evidence for the reef proper.

Notes on Particular Ancient Reefs and Reef-Like Structures

A review of the distribution and biologic components of organic reefs and similar structures then known from the Paleozoic was made in 1911 by Vaughan (pp. 244–45), and an extensive survey of the literature on organic reefs, bioherms, and biostromes was compiled by W. E. Pugh in 1950. Only a very few of the known ancient reefs and reef-like structures have been selected as examples for brief discussion here.

It should be noted that ancient reefs may be in large part essentially structureless. The classic Permian Capitan reef of West Texas and New Mexico is now reasonably well known, especially as a result of the comprehensive areal studies of King (1948) and the succinct and well reasoned interpretive summary of Adams and Frenzel (1950). It is a linear mass of dolomitic and calcitic limestone that, on the one hand, grades with rather sharply interfingering contact to offshore clastic sediments eastward into the Delaware basin, and, on the other hand, passes more gradually to thinly bedded carbonate rocks and evaporites of the back-reef lagoon westward into New Mexico. This reef is a barrier reef, which, once initiated and grown to sea-level, grew mostly upward during or after times of marked depression (Adams and Frenzel, 1950, pp. 297–98, 307). The frame-building organisms presumably were sponges and lime-secreting algae (Adams and Frenzel, 1950, pp. 302, 305; Johnson, 1942). If the general characteristics were not in themselves enough to indicate a wave-breaking structure, the local abundance of dasycladacean algae would be presumptive evidence of shoal depths. As the frame grew its interstices were filled with entrapped detrital materials and presumably also the products of algal and physico-chemical precipitation. This interstitial sediment not only strengthened the frame itself, but became intimately bonded with the frame-builders into a single massive structure. It seems that reef structure has in large part been lost through subsequent though irregular dolomitization. It is estimated by King (1948, p. 85) that the Capitan reef stood 1,000 feet above the basin floor during at least a part of its existence, and Adams and Frenzel (1950, pp. 298, 305, 309–11) give the final relief as "1,200 to more than 1,800 feet." Accumulation of evaporites in the back-reef area indicates sluggishness of circulation in the back-reef lagoon and few passes through the reef.

Similarities between the structure, depositional environment, and dimensions of the Capitan reef and the Visean reef suite of northern England (Bond, 1950) are emphasized in a brief but intriguing paper by Wells (1952). In both instances, the reefs seem to have formed at the margin of a subsiding shelf that surrounded a relatively deep epeiric basin; but the British Visean reefs apparently were individual patch reefs that did not form a continuous structure and presumably indicate growth in relatively quiet water.

The Middle Silurian reefs of the Great Lakes region have been shown by Lowenstam (1950) to have been wave-breaking or potentially wave-resisting structures whose framework was built by stromatoporoids, organisms of unknown affinities, and probably algae. They were, in any event, *bona fide* reefs, as here defined, and one (Marine reef, Lowenstam's Pl. 5) has a shape suggestive of an atoll. The gradational relationships of these particular reefs are discussed in detail by Lowenstam in his classic analysis of this reef suite, and, although differing in many details, they seem to conform with the general scheme of facies relationships here outlined.

The Middle Devonian patch reefs of Michigan have already been briefly discussed, and numerous other fossil patch reefs could be cited. As another example, the Ordovician, Silurian, and Upper Cretaceous reefs of Sweden (Hadding, 1941) appear to be mainly coral, stromatoporoid, and algal patch reefs, and these are at least in part distributed in regular large-scale alignments.

Bioherms of the Upper Cambrian in central Texas (Cloud and Barnes, 1948, Pls. 18, 19A) are masses of compact to very finely granular and thinly laminated stromatolitic limestone that locally shows filamentous structure strongly suggestive of the calcium-precipitating and sediment-binding blue-green algae. It is permissible to retain the name bioherm for structures of this nature because, although they doubtless were formed near the surface, evidence of their potential resistance to strong wave attack is inconclusive. On the other hand, they have the general relationships of patch reefs. The lateral gradation of these bioherms to shale and granular limestone is abrupt. However, at places where it is suspected that local areas of massive dolomite within bedded dolomite may represent similar bioherms, distinctive structures are missing.

One should keep in mind the possibility that some of the "reefs" or "bioherms" of the subsurface may be clastic deposits formed in channels or other depressions of the sea bottom, or mounds or banks of clastic débris that were enabled to accumulate and become indurated as a result of exceptional local factors. Darwin (1889, pp. 277–80) and Branner (1904, pp. 4–200) have described nautical reefs of sandstone that occur intermittently for about 2,000 kilometers off the coast of Brazil (Branner, 1904, p. 5). These are bedded ridges of mostly seaward-dipping quartz sand cemented with calcium carbonate. They presumably originated through induration of beach sands now isolated from the shore. Such "stone reefs," if altogether of calcium carbonate, and if submerged and buried under younger strata, might resemble some of the subsurface "bioherms"; although those of quartz sand, like those of Brazil, would more likely be called "shoestring sands." Lithified dunes, such as those of southeastern Kauai, in the Hawaiian Islands (Hinds, 1930, p. 73; Dana, 1849, p. 277, and 1872, pp. 154–56; personal observations), if they were submerged and buried, would be rather similar to some so-called bioherms—some such dunes on Kauai are elongately dome-shaped, over 40 feet high, and with dips up to 25°. Moreover, as earlier noted, masses of crinoidal débris that are flanked by, and extend abruptly upward through, certain shale successions may well represent concomitant accumulation of dissimilar sediments in adjacent areas—for example, dark muds accumulating on only slightly elevated banks on which lived the crinoids and other shell-bearing organisms, of whose remains the reef-like lenses or wedges are constructed. Some subsurface structures described as reefs may be similar bank deposits even where they are masses of limestone within limestone. Conversely, and as noted earlier, massive *Scolithus* sandstones that have seldom been thought of as reefs may in part be similar to the annelid reefs of the North Sea (Richter, 1927).

Probably few will deny that some so-called reef masses and bioherms now being penetrated by the drill are of uncertain affinities and relationships. More drilling and careful analysis of carefully kept records may eventually provide some of the needed answers, and closely spaced core drilling may be expected to provide more reliable answers than other known subsurface techniques.

REFERENCES

ADAMS, J. E., AND FRENZEL, H. N. (1950), "Capitan Barrier Reef, Texas and New Mexico," *Jour. Geol.*, Vol. 48, pp. 289–312, Fig. 1, Pls. 1–2.

AGASSIZ, ALEXANDER (1903), "The Coral Reefs of the Tropical Pacific," *Mem. Mus. Comp. Zool.*, Vol. 28, 1 vol. text, 410 pp., 238 pls. in 3 separate vols.

BARTON, E. S. (1901), "The Genus Halimeda," *Siboga-Expeditie, Uitkomsten Zool. Bot. Oceanog. en Geol. Gebied*, Mon. 60, pp. 1–32, Pls. 1–4.

BOND, GEOFFREY (1950), "The Lower Carboniferous Reef Limestones of Northern England," *Jour. Geol.*, Vol. 58, pp. 313–29, Figs. 1–3, Pls. 1–2.

BRANNER, J. C. (1904), "The Stone Reefs of Brazil, Their Geological and Geographical Relations, with a Chapter on the Coral Reefs," *Bull. Mus. Comp. Zool.*, Vol. 44 (Geol. Ser., Vol. 7). 285 pp.

CARY, LEWIS R. (1931), "Studies on the Coral Reefs of Tutuila, American Samoa, with Special Reference to the Alcyonaria," Papers from Tortugas Laboratory (*Carnegie Inst. Washington Pub.*), Vol. 27, pp. 53–98, Pls. 1–7, text Figs. 1–14.

CLOUD, P. E., JR. (1942), "Notes on Stromatolites," *Amer. Jour. Sci.*, Vol. 240, pp. 363–79, Fig. 1, Pls. 1–2.

——— (1952), "Preliminary Report on the Geology and Marine Environments of Onotoa Atoll,

Gilbert Islands," *Natl. Res. Council, Pacific Sci. Bd., Scientific Investigations in Micronesia*, Rept. 12. 71 pp.

——— and BARNES, V. E. (1948), "The Ellenburger Group of Central Texas," *Univ. Texas Bur. Econ. Geol. Pub. 4621*. 473 pp.

COUTHOUY, J. P. (1844), "Remarks upon Coral Formations in the Pacific; with Suggestions as to the Causes of Their Absence in the Same Parallels of Latitude on the Coast of South America," *Jour. Boston Soc. Nat. History*, Vol. 4, pp. 66–105, 137–62.

CROSSLAND, CYRIL (1905), "The Ecology and Deposits of the Cape Verde Marine Fauna," *Proc. Zool. Soc. London*, Vol. 1, pp. 170–86, Figs. 21–26.

CULLISON, J. S. (1938), "Origin of Composite and Incomplete Internal Molds and Their Possible Use as Criteria of Structure," *Bull. Geol. Soc. America*, Vol. 49, No. 6, pp. 981–88, Pl. 1.

CUMINGS, E. R. (1932), "Reefs or Bioherms?," *Bull. Geol. Soc. America*, Vol. 43, pp. 331–52.

——— and SHROCK, R. R. (1928), "Niagaran Coral Reefs of Indiana and Adjacent States and Their Stratigraphic Relations," *Bull. Geol. Soc. America*, Vol. 39, pp. 579–620, Figs. 1–12.

DANA, J. D. (1849), "Geology," *U. S. Exploring Expedition 1838–1842*, Vol. 10. 756 pp. and folio atlas of 21 pls.

——— (1872), *Corals and Coral Islands*. London, Sampson Low and Company; New York, Dodd, Mead and Company. 398 pp.

DARWIN, CHARLES (1889), *The Structure and Distribution of Coral Reefs* (3d ed.). London, Smith, Elder and Company. 344 pp.

DAVID, T. W. EDGEWORTH (1904), "Narrative of the Second and Third Expeditions," *Royal Soc. London, Rept. of Coral Reef Comm., The Atoll of Funafuti*, Sec. 4, pp. 40–60, Fig. 17.

———, AND SWEET, G. (1904), "The Geology of Funafuti," *ibid., The Atoll of Funafuti*, Sec. 5, pp. 61–124, Pls. B–E in text, Pls. 1–19 in separate portfolio.

DAVID, T. W. EDGEWORTH, HALLIGAN, G. H., AND FINCKH, A. E. (1904), "Report on Dredging at Funafuti, *ibid.*, Sec. 7, pp. 151–59.

DUNBAR, C. O. (1949), *Historical Geology* (2d ed.). New York, John Wiley and Sons, Inc. 573 pp.

FAIRBRIDGE, R. W. (1950), "Recent and Pleistocene Coral Reefs of Australia," *Jour. Geol.*, Vol. 58, pp. 330–401, Figs. 1–12, Pls. 1–8.

FINCKH, A. E. (1904), "Biology of the Reef-Forming Organisms at Funafuti Atoll," *Royal Soc. London, Rept. of Coral Reef Comm., The Atoll of Funafuti*, Sec. 6, pp. 125–50, Fig. 19.

GARDINER, J. S. (1931), *Coral Reefs and Atolls*. London, Macmillan and Company, Ltd. 181 pp.

HADDING, ASSAR (1941), "The Pre-Quaternary Sedimentary Rocks of Sweden, VI, Reef Limestones," *Lunds Universitets Arsskrift*, N. F., Avd. 2, Bd. 37, Nr. 10. 137 pp.

——— (1950), "Silurian Reefs of Gotland," *Jour. Geol.*, Vol. 58, pp. 402–09, Figs. 1–3, Pl. 1.

HARTT, C. F. (1870), *Geology and Physical Geography of Brazil*. Boston, Fields, Osgood, and Company. 620 pp.

HENSON, F. R. S. (1950), "Cretaceous and Tertiary Reef Formations and Associated Sediments in Middle East," *Bull. Amer. Assoc. Petrol. Geol.*, Vol. 34, No. 2, pp. 215–38, Figs. 1–14.

HINDE, G. J. (1904), "Report on the Materials from the Borings at the Funafuti Atoll," *Royal Soc. London, Rept. of Coral Reef Comm., The Atoll of Funafuti*, Sec. 11, pp. 186–361, Figs. 21–22.

HINDS, N. E. A. (1930), "The Geology of Kauai and Niihau," *Bernice P. Bishop Mus. Bull. 71*. 103 pp.

HOFFMEISTER, J. E. (1932), "Geology of Eua, Tonga," *ibid., Bull. 96*. 93 pp.

JOHNSON, J. H. (1942), "Permian Lime-Secreting Algae from the Guadalupe Mountains, New Mexico," *Bull. Geol. Soc. America*, Vol. 53, pp. 195–226, Figs. 1–5, Pls. 1–7.

——— (1943), "Geologic Importance of Calcareous Algae with Annotated Bibliography," *Quar. Colorado School of Mines*, Vol. 38, No. 1. 102 pp.

——— (1946), "Late Paleozoic Algae of North America," *American Midland Naturalist*, Vol. 36, No. 2, pp. 264–74, Pls. 1–2.

KING, P. B. (1948), "Geology of the Southern Guadalupe Mountains," *U. S. Geol. Survey Prof. Paper 215*. 183 pp.

KREMPF, ARMAND (1927), "La forme des récifs corallien et le régime vents alternants," *Travaux du Service Oceanographique des Pêches de l'Indochine*, Mém. 2, pp. 3–33, Figs. 1–4, Pls. 1–6.

KUENEN, P. H. (1933), "Geology of Coral Reefs," *The Snellius Exped. (Eastern Neth. E. Ind. 1929–30)*, Vol. 5 (Geol. Results), Pt. 2. 125 pp. Kemink en Zoon, N. V. Utrecht.

——— (1947), "The Borings on Maratoea and the Coral Reef Theory of Glacially Controlled Subsidence," *Koninklijke Ned. Akad. Wetenschap. Afd. Natuurkunde, Verhandelingen*, Tweede Sectie, Deel 43, No. 3, pp. 1–35, Figs. 1–14, Pls. 1–3.

LADD, H. S. (1944), "Reefs and Other Bioherms," *Natl. Research Council Comm. on Marine Ecol. as Related to Paleont.*, Rept. 4, pp. 26–29.

LADD, H. S., TRACEY, J. I., JR., AND LILL, G. G. (1948), "Drilling on Bikini Atoll, Marshall Islands," *Science*, Vol. 107, No. 2768, pp. 51–55, Figs. 1–2.

LADD, H. S., TRACEY, J. I., JR., WELLS, J. W., AND EMERY, K. O. (1950), "Organic Growth and Sedimentation on an Atoll," *Jour. Geol.*, Vol. 58, No. 4, pp. 410–25, Figs. 1–2, Pls. 1–7.

LECOMPTE, MARIUS (1938), "Quelques types de 'recifs' Siluriens et Dévoniens de l'Amerique du Nord. Essai de comparison avec les récifs coralliens actuels," *Musée Royal d'Histoire Naturelle de Belgique*, tome 14, no. 39. 51 pp.

——— (1951), "Les stromatoporoides du Dévonien moyenet supérieur du bassin de Dinant (Pt. 1)," *Institut Royal des Sciences Naturelles de Belgique*, Mem. 116. 215 pp.

LEMOINE, MME. PAUL (1940), "Les algues calcaires de la zone néritique," from "Contribution a l'étude de la répartition actuelle et passée des organismes dans la zone néritique," *Soc. de Biogéographie*, 7, pp. 75–138. Paris, Paul Lechevalier.

LOWENSTAM, H. A. (1950), "Niagaran Reefs of the Great Lakes Area," *Jour. Geol.*, Vol. 58, No. 4, pp. 430–87, Figs. 1–11, Pls. 1–5.

NEWELL, NORMAN D., ET AL. (1951), "Shoal-Water Geology and Environments, Eastern Andros Island, Bahamas," *Bull. Amer. Mus. Nat. History*, Vol. 97, Art. 1, pp. 1–29, Figs. 1–5, Pls. 1–8.

PIA, JULIUS (1927), "Thallophyta," Bd. 1, Abt. 1, pp. 31–136, 693–700, in M. HIRMER, *Handbuch der Paläobotanik*. Munchen and Berlin.

——— (1933), "Die rezenten Kalksteine," *Zeitschr. Kristallogr. Miner. und Petrogr.*, Abt. B. Miner. und Petrogr. Mitteil. Ergänzungsband. 420 pp.

——— (1937), "Die wichtigsten Kalkalgen des Jungpaläozoikums und ihre geologische Bedeutung," *Compte Rendu deuxieme Congrês pour l'avancement des études de Stratigraphie Carbonifère* (Heerlen, 1935), pp. 765–856, Fig. 1–2, Pls. 85–97.

PUGH, W. E. (1950), *Bibliography of Organic Reefs, Bioherms, and Biostromes*. Tulsa, Oklahoma, Seismograph Service Corporation. 139 pp.

QUOY, J. R. C., AND GAIMARD, J. P. (1824), *Voyage autour du monde (Freycinet; 1817–1820), Zoologie*. Paris, Pillet Aîné. 712 pp.

RICHTER, RUDOLPH (1927), " 'Sandkorallen'-Riffe in der Nordsee," *Senckenbergische Naturf. Gesellschaft*, Natur u. Museum, Bd. 57, Heft 2, pp. 49–70.

SCHULTZ, L. P. (1943), "Fishes of the Phoenix and Samoan Islands Collected in 1939 during the Expedition of the U. S. S. *Bushnell*," *U. S. Nat. Mus. Bull. 180*. 316 pp.

SETCHELL, W. A. (1928), "A Botanical View of Coral Reefs, especially of Those of the Indo-Pacific Region," *Proc. Third Pan-Pac. Cong.*, Vol. 2, pp. 1837–43.

SHEPARD, F. P. (1948a), "Evidence of World-Wide Submergence," *Sears Found., Jour. Marine Research*, Vol. 7, No. 3, pp. 661–78, Figs. 1–6.

SMITH, G. M. (1938), *Cryptogamic Botany, Vol. 1, Algae and Fungi*. New York and London, McGraw-Hill and Company. 545 pp.

———, ET AL. (1951), *Manual of Phycology*. Waltham, Massachusetts, Chronica Botanica Company. 375 pp.

SOLLAS, W. J. (1904), "Narrative of the Expedition in 1896," *Royal Soc. London, Rept. of Coral Reef Comm., The Atoll of Funafuti*, Sec. 1, pp. 1–28, Figs. 1–16.

TAYAMA, RISABURO (1935), "Table Reefs, a Particular Type of Coral Reefs," *Proc. Imper. Acad. Tokyo*, Vol. 2, No. 7, pp. 268–70.

UMBGROVE, J. H. F. (1929), "De Koraalriffen der Duizend-Eilanden (Javazee)," *Dienst van dem Mijnbouw in Nederlandsch-Indië, Wetensch. Mededeel.*, No. 12, pp. 1–47, Figs. 1–11, Pls. 1–6.

——— (1946), "Corals from a Lower Pliocene Patch Reef in Central Java," *Jour. Paleon.*, Vol. 20, No. 6, pp. 521–42, Pls. 77–82.

——— (1947), "Coral Reefs of the East Indies," *Bull. Geol. Soc. America*, Vol. 58, pp. 729–78, Figs. 1–24, Pls. 1–7.

——— (1949), *Structural History of the East Indies*. Cambridge, England, Cambridge Univ. Press. 63 pp.

VAUGHAN, T. W. (1911), "Physical Conditions under Which Paleozoic Coral Reefs Were Formed," *Bull. Geol. Soc. America*, Vol. 22, pp. 238–52.

——— (1919), "Corals and the Formation of Coral Reefs," *Smithsonian Inst. Ann. Rept., 1917*, pp. 189–276, Figs. 1–16, Pls. 1–37.

———, AND WELLS, J. W. (1943), "Revision of the Suborders, Families, and Genera of the Scleractinia," *Geol. Soc. America Spec. Paper 44*. 363 pp.

VERWEY, J. (1931a), "The Depth of Coral Reefs in Relation to Their Oxygen Consumption and the Penetration of Light in the Water," *Treubia*, Pt. 2, Vol. 13, pp. 169–98.

——— (1931b), "Geomorphological Notes on the Coral Reefs of Batavia Bay," *ibid.*, Vol. 13, Livr. 2, pp. 199–216, Figs. 1–4, Pls. 1–2.

WEBER VAN BOSSE, MME. A., AND FOSLIE, M. (1904), "The Corallinaceae of the Siboga-Expedition," *Siboga-Expeditie, Uitkomsten Zool. Bot. Oceanog. en Geol. Gebied*, Mon. 61. 110 pp.

WELLS, J. W. (1952), "Note on Mississippian and Permian Reef Suites," *Jour. Geol.*, Vol. 60, No. 1, pp. 97–98, Fig. 1.

Reprinted from
BULLETIN OF THE AMERICAN ASSOCIATION OF PETROLEUM GEOLOGISTS
VOL. 37, NO. 5 (MAY, 1953), PP. 1014-1029, 4 FIGS., 1 PLATE, 2 TABLES

PETROGRAPHY AND PETROLOGY OF SCURRY REEF, SCURRY COUNTY, TEXAS[1]

RICHARD E. BERGENBACK AND ROBERT T. TERRIERE[2]
Columbus, Ohio, and Austin, Texas

ABSTRACT

The Scurry reef of Scurry County, Texas, is a subsurface feature 23 miles long and 4–8 miles wide that is of Permian and Carboniferous age. The reef is a low mound composed chiefly of clastic organic limestone with minor amounts of black bituminous claystone.

The reef limestone consists of three main textural types: calcilutite, calcarenite, and calcirudite. The calcilutite is composed mainly of particles too fine to be resolved under the petrographic microscope. Part of this fine material appears to be comminuted organic débris; the remainder may be chemically precipitated ooze. The calcarenite forms a gradational series between sorted, rounded débris in clear crystalline calcite cement and unsorted, angular débris in a matrix of lithified lime mud. The calcirudite is composed of angular, granule- to boulder-sized fragments in a matrix of lithified lime mud or lime sand. The fragments consist of organic débris and pre-existing reef limestone and claystone.

The limestone contains minor amounts of secondary calcite, dolomite, and chert. Characteristic structures in the limestone are stylolites, joints, irregular fractures, and slump structures.

The vast bulk of observable porosity was formed by leaching solutions. This condition may have been accomplished principally by the enlargement of primary openings through the removal of the least stable constituents.

The Scurry reef was formed from the hard parts of organisms that lived, died, and were buried in essentially one locality. The two major faunal elements that could be identified in the reef are the Crinoidea and the Foraminifera, principally Fusulinidae. Corals constitute a minor element and are not numerous enough to be important here as reef-forming organisms.

It is suggested that the welding of lime mud and the constant precipitation of calcite cement rigidified the accumulating material and thus acted as a substitute for the rigid organic framework that characterizes many reefs.

INTRODUCTION

The Scurry reef of Scurry County, Texas, is a subsurface feature 23 miles long and 4 to 8 miles wide that includes the Diamond-M and Kelly-Snyder oil-producing areas (Figs. 1 and 2). The reef is a low mound composed of clastic fragments of marine organisms. It is flanked on all sides by black claystone and contains lenticular black claystone beds. The irregular top of the reef is 3,738 ft. to more than 4,700 ft. below sea level. In a recent study (Heck, Yenne, and Henbest, 1952) the Scurry reef was assigned to the Permian and Carboniferous periods.

Petrographic studies of the Scurry reef were made to determine the composition and texture of the reef limestone and the diagenetic changes that it has undergone.

Work was begun in April, 1951, as a part of a reservoir study of the reef made by the Fuels Branch of the U. S. Geological Survey working in cooperation with the U. S. Bureau of Mines under the auspices of the Petroleum Administration for Defense. The work was directed by Howard E. Rothrock, Fuels Branch, U. S. Geological Survey. Paul D. Krynine examined 30 representative thin sections and contributed numerous excellent observations and interpretations.

[1] Manuscript received, September 29, 1952. Publication authorized by the director of the United States Geological Survey, Washington, D. C.

United States Geological Survey.

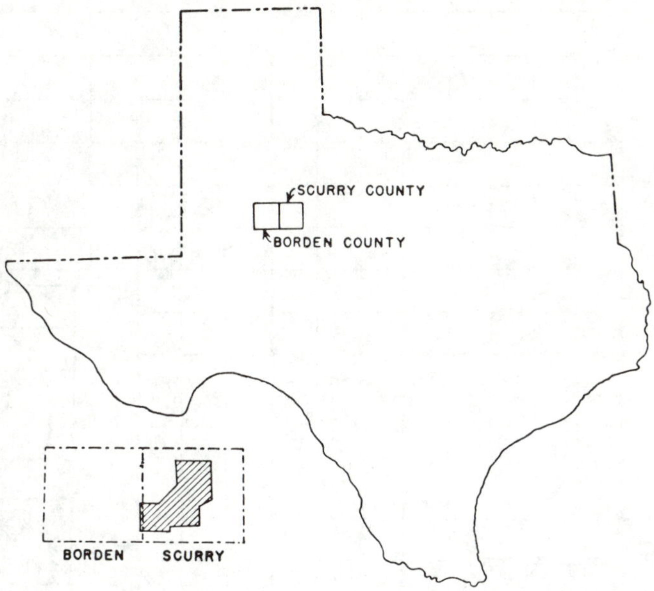

FIG. 1.—Index map of Texas showing area covered by this report.

METHOD

The initial stages of the investigation included a complete megascopic examination of 4,734 ft. of core from 73 wells in the reef. The cores are 2 to 4 in. in diameter and each core represents 10 to 300 ft. of stratigraphic section. The locations of the cores and the series numbers assigned to them are shown in Fig. 2.

Freshly broken, wetted core surfaces were megascopically examined with the aid of a hand lens or binocular microscope. The features observed and tabulated comprise rock type, bedding, color, texture, general type and abundance of fossils, structural features, secondary mineralization, and type of porosity.

Samples for thin-section study were taken so that the various textures and features of special interest in the reef limestone could be examined. Thin sections were made from 340 of these samples.

REEF LIMESTONE

The Scurry reef consists chiefly of gray clastic organic limestone and minor amounts of black bituminous claystone.

Distinct bedding planes were not observed in the limestone because the various limestone textures grade irregularly into each other. Bedding of the claystone is irregular and some claystone is transitional with adjacent limestone. Some very thin claystone stringers dip as much as 30°. These relatively high dips and the irregularities in the claystone beds probably are due to slumping.

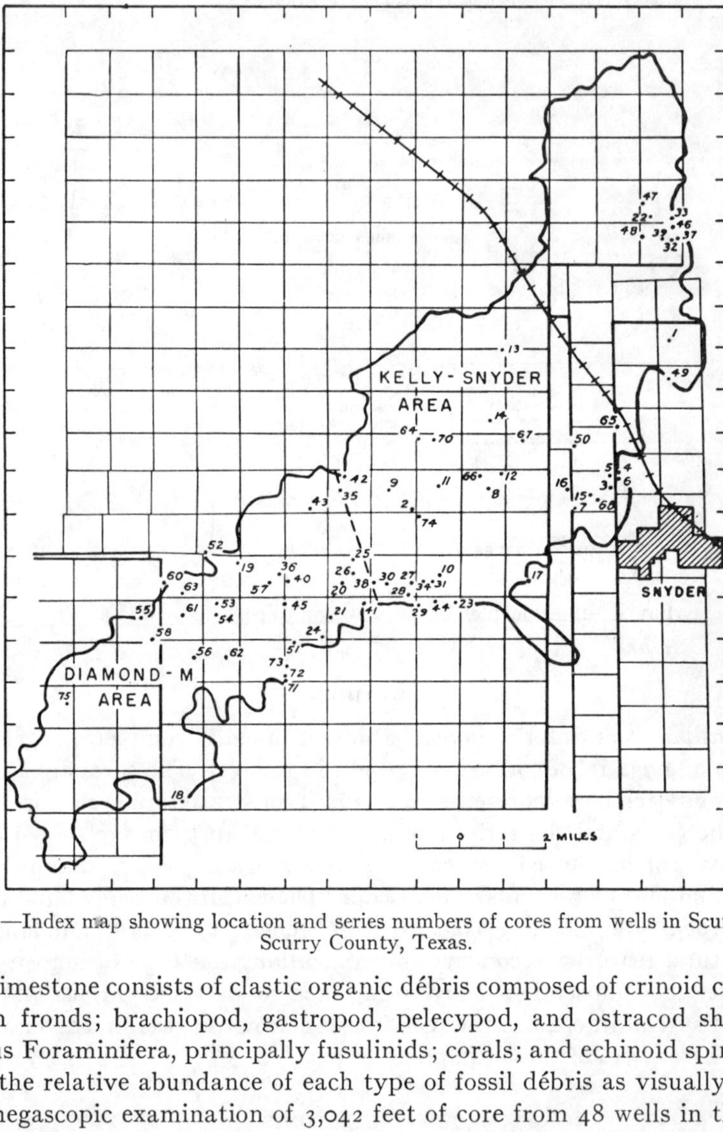

FIG. 2.—Index map showing location and series numbers of cores from wells in Scurry reef, Scurry County, Texas.

The limestone consists of clastic organic débris composed of crinoid columnals; bryozoan fronds; brachiopod, gastropod, pelecypod, and ostracod shells; tests of various Foraminifera, principally fusulinids; corals; and echinoid spines. Table I shows the relative abundance of each type of fossil débris as visually observed during megascopic examination of 3,042 feet of core from 48 wells in the Scurry reef.

The reported high percentage of brachiopods and low percentages of foraminiferal (other than fusulinid) débris are questionable, because of a tendency to unconsciously overemphasize large fragments when reporting the amount of each type of fossil débris and to overlook many of the smaller Foraminifera.

It was convenient to use the textural terms calcarenite, calcilutite, and calcirudite for the clastic limestone. Pettijohn (1949, pp. 300, 307) defines them as follows.

Calcarenite is a general term to describe those detrital carbonate rocks of sand grain size (1/16–2 mm. in diameter) that are with or without calcite cement and that are composed mainly (over 50 per cent) of carbonate detritus. If the fragments are over 2 mm. in diameter, the term calcirudite may be applied to the rock. . . . By decrease in grain size, the calcarenites grade into the calcilutites.

CALCILUTITE

The calcilutite of the reef, although composed mainly of particles too fine to recognize megascopically, commonly contains varying amounts of unsorted calcareous organic débris coarser than $\frac{1}{16}$ mm. in diameter.

Microscopic examination failed to reveal the origin of many of the carbonate particles since most of them are submicroscopic (Pl. 1, A). Some of the visible particles of the calcilutite consist of comminuted fossil débris that represents an end product of attrition. The remainder may be chemically precipitated ooze.

TABLE I. APPROXIMATE ABUNDANCE OF INVERTEBRATE FOSSIL TYPES IN 3,042 FEET OF CORE FROM SCURRY REEF, SCURRY COUNTY, TEXAS

(Data compiled by Donald A. Myers, United States Geological Survey)

Fossil Type	*Percentage of 3,042 Feet of Core Containing Fossil Varieties*				
	Rare	*Occasional to Common*	*Very Common to Abundant*	*Present (Abundance not Recorded)*	*Total*
Echinoderms	20	51	6	4	81
Fusulinids	26	39	7	2	74
Brachiopods	24	14	—	9	47?
Bryozoans	13	17	—	6	36
Corals	7	1	—	4	12
Foraminifera (other than fusulinids)	3	6	—	2	11?
Mollusks	6	2	—	1	9

The comminution of some of the débris may have been accomplished by passage through the digestive tracts of scavenging organisms. Some of the very fine-grained material contains dark ovoid bodies, smaller than 0.05 mm. in diameter, which may represent fecal pellets of such organisms. Investigators of recent reefs (Darwin, p. 21) and other recent sediments (Twenhofel, 1939, p. 77) have pointed out that worms and other scavengers are important in forming lime mud.

Chemical precipitation of lime ooze may have been related to organic agents such as bacteria or algae, but no evidence of such agents was found. Much chemically precipitated calcium carbonate is aragonite rather than calcite (Pettijohn, 1949, p. 108), but it can not be determined whether or not the lime muds of the Scurry reef originally contained aragonite crystals.

Most of the fragments coarser than $\frac{1}{16}$ mm. are crinoid columnals, bryozoan fronds, shell fragments, and tests of Foraminifera. Other coarse fragments that have outlines closely resembling organic débris consist of clear anhedral calcite mosaics (Pl. 1, B). These latter fragments are interpreted to be pieces of organic débris that have been reorganized to, or replaced by, crystalline calcite.

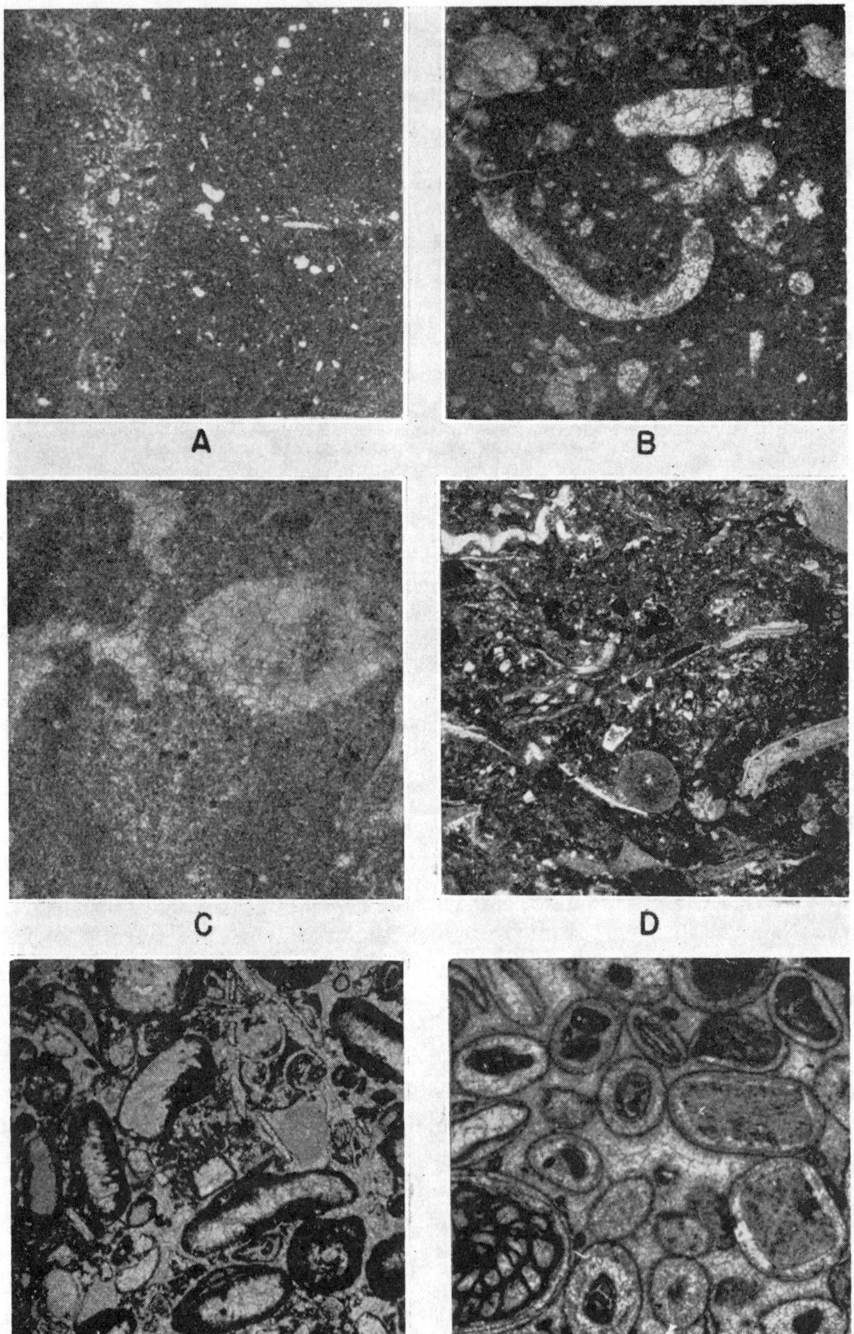

PLATE I

The calcilutite contains scattered areas of anhedral calcite (Pl. 1, C). The crystalline material probably is mostly primary chemical cement but may have been formed by reorganization or replacement of fine aragonite or calcite mud.

CALCARENITE

The calcarenite of the Scurry reef is composed mainly of unsorted to well sorted, sharply angular to well rounded organic fragments that range in diameter from $\frac{1}{16}$ to 2 mm. Reef rock fragments, some coarser than 2 mm., and oölites and pisolites, are present in small amounts. Fragments composed of mosaics of clear crystalline calcite, similar to those in the calcilutite, are common. A special variety of calcarenite—or calcirudite, depending on grain size—that is composed almost entirely of crinoid columnal fragments is termed encrinite (Pettijohn, 1949, p. 301).

The calcarenite of the Scurry reef may be considered a gradational series between two extremes. One extreme is calcarenite composed of partly sorted to well sorted, subrounded to rounded organic débris in a clear crystalline calcite cement (Pl. 1, E). The other is calcarenite composed of unsorted angular débris in a matrix of lithified lime mud (Pl. 1, D). Most of the calcarenite is intermediate between the two extremes and exhibits properties of both, but it resembles more closely the unsorted extreme. A special variety of the sorted extreme is composed of oölites [illegible] crystalline calcite cement (Pl. 1, F).

The lithified lime-mud matrix of the unsorted extreme is identical with the material composing the calcilutite. As the amount of matrix increases, the unsorted calcarenite [illegible] the dividing line between the two being drawn where [illegible] coarser than $\frac{1}{16}$ mm.

The [illegible] sorted calcarenite might be interpreted either as [illegible] as primary chemical cement. The [illegible] is a primary chemical precipitate for the following [illegible] cement is present in considerable [illegible] relatively high degree of sorting and rounding. This suggests that much of the interstitial fine material was winnowed out. This winnowing process would create interstitial voids favorable for the [illegible] (b) In most calcarenite containing [illegible] boundaries are coated with a fine

---

[illegible] URES OF LIMESTONE OF SCURRY REEF

[illegible] probably replaced or reorganized organic débris.

[illegible] may be product of replacement or reorganiza- [illegible]

D. [illegible] organic débris in matrix of lime mud. Ordinary light, ×7.5.

E. Calcarenite composed of partly sorted, rounded, organic débris in crystalline cement. Ordinary light, ×12.

F. Accretion-type oölites in matrix of crystalline calcite cement. Lime-mud layers in oölites have been largely reorganized or replaced by crystalline calcite. Ordinary light, ×15.

drusy fringe of calcite crystals. This coating suggests primary interstitial void filling. (c) Prominent crystalline calcite overgrowths on many grains, especially on crinoid columnals, are in optical continuity. According to Pettijohn (1949, p. 305), this continuity is a characteristic of primary cement. (d) There are no relict structures of grains or fossils in the crystalline cement as would be expected if the cement were the product of a reorganized or replaced matrix. (e) A few reef-rock fragments of calcilutite are present as unaltered grains in calcarenite that has crystalline cement. It seems improbable that reorganization of interstitial lime mud to crystalline calcite would take place without affecting a fragment of the same material. This reasoning might not be valid if the reorganization of the matrix were considered to have taken place before lithification. (f) In a few specimens of calcarenite, grains cemented by crystalline calcite are coated with a thin layer of lime mud that apparently adhered to the grains during deposition. These lime-mud coatings show no evidence of reorganization. (g) Some calcite-filled fusulinid tests are cut by fractures that do not continue into the matrix (Fig. 3). The presence in the fusulinid tests of crystalline calcite that is obviously older than the lithification of the matrix indicates that conditions were favorable for the precipitation of primary calcite cement.

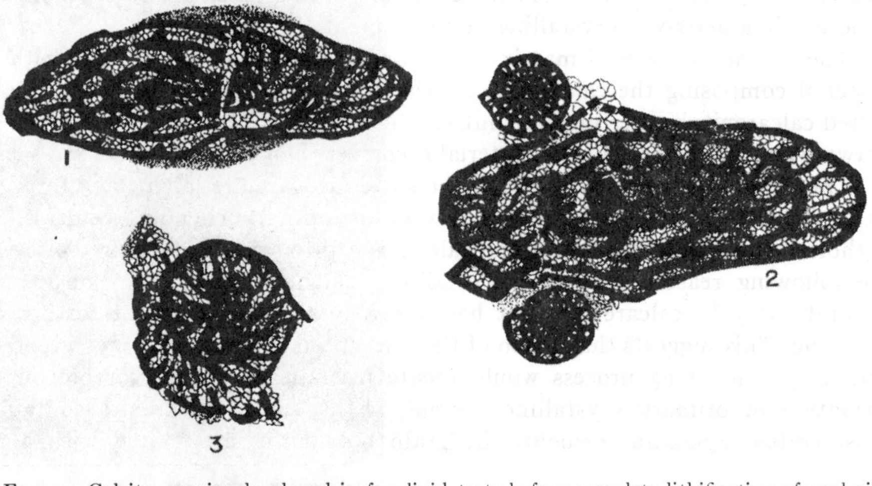

FIG. 3.—Calcite mosaics developed in fusulinid tests before complete lithification of enclosing matrix.

1. Developmental stages
   a. Formation of fine mosaic in empty fusulinid chambers
   b. Fracturing of filled tests before complete lithification of matrix
   c. Filling of fracture by coarser calcite
2. Developmental stages
   a. Crushing of empty test by compaction between adjacent more rigid forms before lithification
   b. Formation of mosaic in empty chambers
   c. Continued growth of mosaic causing expansion of broken margin of test
3. Anhedral calcite filling that has broken fusulinid test by expansive force of crystallization

Magnification ×16.

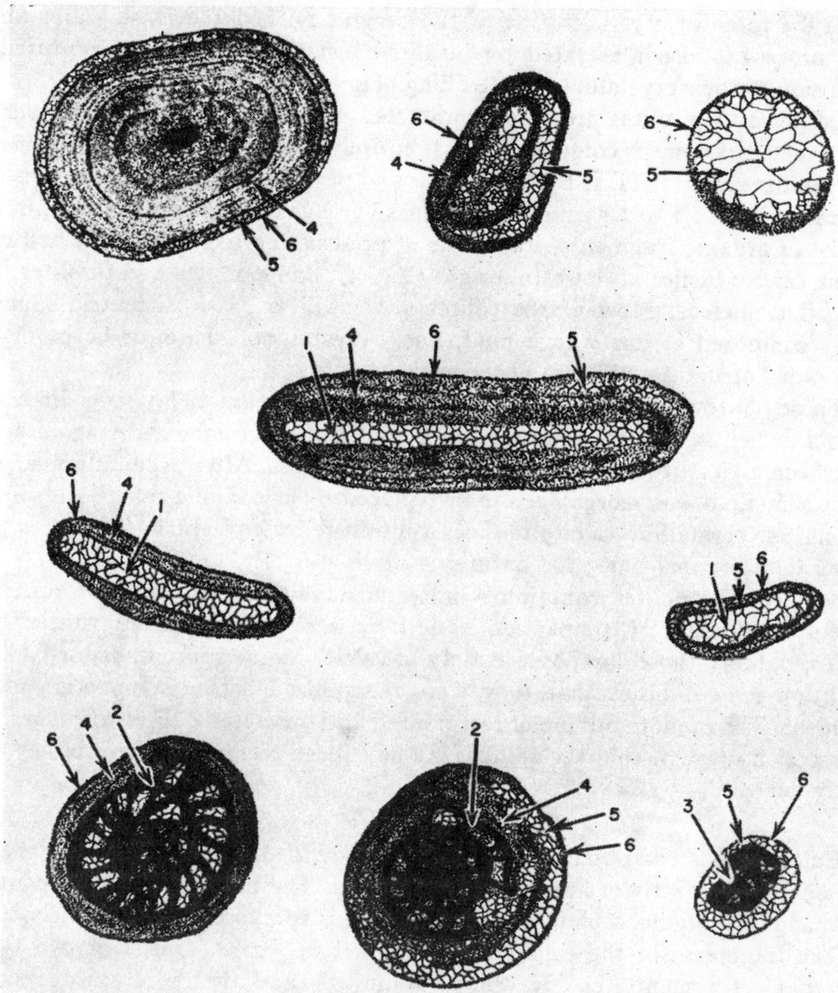

FIG. 4.—Accretionary calcareous oölites and pisolites showing different nuclei and various stages of calcite reorganization.

1. Nucleus consisting of crystalline calcite fragment
2. Nucleus consisting of calcite-filled fusulinid
3. Nucleus of pre-existing limestone
4. Dusty lime-mud layers accumulated by accretion
5. Anhedral calcite mosaic formed by reorganization of lime mud
6. Surface coating of lime mud

Magnification ×17.

These observations indicate that an important factor in the lithification of calcarenite was cementation by crystalline calcite and that this process took place early in the history of the reef.

Some grains cemented by crystalline calcite do not touch, presumably because the force of crystallization of the cement pushed the loose grains apart. This process is also illustrated by fusulinid tests that have been ruptured by expansion of their crystalline calcite filling (Fig. 3).

Oölites and pisolites are rare in unsorted calcarenite, but a special variety of the sorted extreme is composed almost entirely of well sorted oölites in a matrix of crystalline calcite (Pl. 1, F). The oölites and pisolites are spherical to ellipsoidal and range from 0.2 to 2.5 mm. in diameter. The nuclei of the oölites and pisolites consist of organic fragments, fragments of pre-existing reef limestone, and crystalline calcite bodies of doubtful origin (Fig. 4). Some oölites and pisolites have no visible nucleus. Most of the oölites and pisolites have concentric bands or layers composed of dusty lime mud, finely crystalline calcite, or both. Oölites with radial structure were not observed.

These oölites were probably formed by accumulation of layers of lime mud around nuclei as they rolled around in a lime-mud environment. Aragonite may have been a constituent of the encrusting mud layers. After accumulation, some of the lime mud was reorganized to or replaced by crystalline calcite. In some of the oölites, crystalline calcite mosaics contain relict concentric structures indicating their original lime-mud nature.

Some of the oölites containing lime-mud layers are squashed or ruptured, possibly as a result of compaction while they were still in a plastic state. Other oölites, notably those composed mainly of calcite mosaics, are undistorted. This condition may indicate that they were reorganized before compaction of the sediment. The random mixing of reorganized and unaltered oölites within a single thin section suggests that the oölites may have been reorganized before they were finally deposited.

CALCIRUDITE

Calcirudite is composed of angular to subrounded, granule- to boulder-sized fragments in a matrix of lime sand or lime mud. The matrix commonly contains some admixed argillaceous material but rarely is composed entirely of clay.

The fragments of the calcirudite coarser than 2 mm. consist of two types: (a) organic fragments, chiefly crinoid columnals and the more sturdy part of shells; (b) fragments of pre-existing reef limestone and claystone.

Coarse calcirudite that consists predominantly of pre-existing reef limestone is termed reef breccia. The largest fragment of pre-existing reef limestone measured is 1.6 feet in diameter. Fragments of this size may be more common than has been reported, because the megascopic descriptions were made from short sections of core, with a maximum diameter of 4 inches, so that it is not always possible to differentiate large clastic blocks from calcarenite and calcilutite lenses. The pieces of pre-existing reef limestone are identical with the unbroken calcarenite and calcilutite in the reef. Fragments of calcarenite and calcilutite are mixed indiscriminately in a matrix of either lime sand or lime mud.

Many contacts between adjoining fragments in the calcirudite are stylolitic, but most stylolites do not continue into the matrix. This discontinuity indicates that the stylolites developed between the fragments, presumably by pressure solution, before the matrix lithified.

The fragments of older reef limestone are generally angular and undoubtedly were lithified before they were broken and redeposited. Their presence within the reef mass shows that the reef was continuously lithified as it grew.

POROSITY

The porosity of the limestone of the Scurry reef, as observed megascopically, ranges from almost zero to an estimated 20 per cent and is developed in irregularly spaced zones of highly variable thickness. The visible openings range in size from tiny pin-point openings to vugs 5 cm. in diameter, but most of the openings are smaller than 3 mm. in diameter. Open joints and fissures contribute an undetermined additional amount of porosity.

The megascopic porosity appears to be secondary and to have developed irrespective of grain size. In some places leaching has been confined almost entirely to the interstitial filling, but elsewhere it has affected the grains more than the filling. Fragments of echinoderms, recent forms of which are reported by Clarke and Wheeler (1917, pp. 20–30) to contain a relatively high percentage of magnesium carbonate, are notably unaffected by leaching. In most cases, however, it is impossible to assign a definite reason for the leaching of one constituent and not another.

Examination under the microscope reveals a relatively insignificant amount of primary porosity in the reef limestone. Hollow chambers of fusulinid tests, lined with only a thin fringe of drusy calcite, are the most obvious example of primary porosity in this rock. Some calcilutite and the lime-mud matrix of calcarenite contain very minute voids, probably because of incomplete welding of lime mud.

Most of the porosity is secondary. The leaching that produced this secondary porosity may have been active principally in enlarging primary openings and in removing the least stable constituents, such as aragonite.

Deposition of secondary carbonate has decreased the amount of porosity and locally has completely closed some of the voids.

SECONDARY MINERALIZATION

Calcitization.—Calcitization appears to have begun early in the history of the reef and continued, probably with interruptions, into much later stages. In addition to the primary cement and reorganized mosaics of calcite already mentioned, calcite occurs as drusy scalenohedrons and rhombohedrons lining vugs, open joints, and fractures. In places, crystalline carbonate has completely filled the openings.

Dolomitization.—Dolomite was observed megascopically in several cores in the form of drusy rhombohedrons lining vugs, and as disseminated rhombohedrons. Microscopic examination revealed more dolomite than had previously been suspected. Well formed rhombohedrons of dolomite, many of which are slightly brownish, are scattered through some of the calcilutite and the lime-mud matrix of some of the calcarenite. Dolomite rhombohedrons are extremely rare in the clastic fragments of the reef. Paul D. Krynine, United States Geological Survey consultant, pointed out that the dolomite occurs also as an ill defined, finely crystalline mosaic in the lime mud of several thin sections.

Krynine believes that primary porosity was very important in the control of dolomitization and that much of the dolomite replacement occurred very shortly after deposition of the limestone. Dolomitization probably began with the replacement of lime-mud matrix by euhedral rhombohedrons, followed by the development of ill defined dolomite mosaics as replacement continued. Further dolomitization replaced the fine-grained parts of clastic fragments and finally even affected some of the coarser-grained elements. In only a very few places, however, did dolomitization continue to the extent that entire clastic fragments were affected.

Silicification.—Small amounts of chert were observed megascopically in 4 of the 73 cores examined. The chert occurs as gray to white layers 4–9 in. thick and as irregular masses as much as 5 cm. in diameter. The edges of the chert layers are generally irregular. The most distinctive of these irregular boundaries is a stylolite of high amplitude between white chert and dense calcilutite. This boundary is evidence that the chert is secondary and was formed after the stylolite. The pressure solution process by which stylolites are thought to be formed could not have produced so serrate a contact between rocks of such dissimilar hardness and solubility as chert and limestone.

Chert was seen under the microscope as pale yellowish brown microcrystalline granular and fibrous aggregates. The granular chert appears as patches replacing fossil débris, commonly crinoid columnals, and rarely replacing interstitial fill. Fibrous microcrystalline silica is much less common and is usually found as a lining in vugs.

Evidence indicates that silicification occurred during two different stages of reef history. Part of the silicification occurred before the reef was dolomitized, as shown by areas in which silicification has proceeded randomly across areas of calcite and dolomite. The replacement probably would have been selective if both dolomite and calcite had been present, because of the greater solubility of calcite. In one thin section, differential leaching has caused a knob of chert to project into an otherwise ovate vug, showing that silicification in that part of the reef must have taken place prior to leaching.

A second period of silicification was characterized by the deposition of secondary crystalline quartz in vugs and fractures. A small amount of chert also seems to have been deposited at this time. In at least one thin section, microcrystalline granular silica that has replaced carbonate grades almost imperceptibly into a clear crystalline quartz mosaic filling a vug. In the same thin section microcrystalline fibrous silica appears as a lining of vugs, the centers of which are filled with crystalline quartz.

Other mineralization.—Tiny, almost submicroscopic, crystals concentrated in bituminous material were identified by Krynine as feldspar, probably albite.

Collophane occurs in a few of the thin sections. Some of it appears to have replaced organic remnants, but part of it may be fragments of phosphatic shells.

Pyrite is present in some of the claystone and occurs sporadically as isolated cubes or pyritohedrons in limestone of the Scurry reef.

CLAYSTONE

The claystone of the Scurry reef is dark gray to black, because of a high percentage of bituminous material. It contains a considerable amount of unsorted carbonate débris.

Many thin bituminous claystone stringers within the reef pass laterally into stylolites containing bituminous layers as much as 3 mm. thick. Megascopically it is commonly difficult to differentiate thin claystone stringers from dried bituminous residue along solution zones, because of the large amount of admixed bituminous material in the claystone.

Under the microscope the claystone appears as a black to brownish mass containing a large amount of clastic limestone particles. The edges of some of these particles of limestone are ragged, probably because of solution. Some of the limestone fragments have been silicified. Chert also occurs as microscopic irregular masses in the claystone.

The dark-colored matrix of the claystone is composed of a mixture of black, opaque bituminous material and brown, translucent isotropic material, part of which may be collophane.

Pyrite, very minute particles of quartz and feldspar, and small flakes of detrital muscovite are present in small amounts. The pyrite occurs as pyritized spicules of unknown origin averaging 1 mm. in diameter and as much as 5 cm. in length; disseminated cubes and pyritohedrons; nodules as large as 2 cm. in diameter; and, near the top of the reef, pyritized plant fragments and brachiopod shells. The actual percentage of pyrite in the claystone is very small.

DISTRIBUTION AND ABUNDANCE OF TEXTURAL VARIETIES

Calcarenite is the most abundant of the textural varieties described from the Scurry reef cores. Calcilutite, calcirudite, and claystone are next in order of abundance (Table II).

The interlensing of the three textural types of limestone is complex. The average thickness of the textural units reported in the megascopic descriptions is 4 ft. This figure is somewhat greater than the actual average thickness because many thin layers were not described separately.

Coarse calcirudite, claystone, and oölitic calcarenite tend to be concentrated

TABLE II. TOTAL FOOTAGE AND RELATIVE ABUNDANCE OF REEF TEXTURES AS DESCRIBED FROM SCURRY REEF CORES

Textural Types	*Total Thickness Described (Feet)*	*Relative Abundance (Percentage)*	*Average Thickness of Described Units (Feet)*	*Range in Thickness of Described Units (Feet)*
Calcarenite	2,339	49.4	6.7	Max. 79.5 Min. 0.3
Calcilutite	1,676	35.4	6.1	Max. 41.0 Min. 0.2
Calcirudite	638	13.4	4.7	Max. 26.0 Min. 0.3
Claystone	81	1.8	2.8	Max. 8.5 Min. 0.2

along the flanks of the reef and are mostly within a relatively non-porous depth zone called the intermediate zone (Rothrock *et al.*, *in preparation*). The other textural types of reef limestone apparently are randomly distributed.

STRUCTURAL FEATURES

The characteristic structures in the reef limestone are stylolites, joints, irregular fractures, and slump structures.

STYLOLITES

Incipient to well developed stylolites are common in the reef limestone. Some of them are extremely serrate, with amplitudes of as much as 4 in. Stylolites are especially common between adjoining fragments in the calcirudite. The presence of thoroughly leached limestone and unleached limestone on opposite sides of some of these stylolites suggests that they formed before leaching.

In addition to typical serrate stylolites, the limestone of the Scurry reef contains similar but more nearly linear features that seem to have been formed by leaching along horizontal fractures. These solution features contain considerable bituminous matter, as do the normal stylolites. Some of these features contain microscopic lens-shaped and sharply angular limestone fragments with etched boundaries.

JOINTS

Joints are present in most of the core. They are generally vertical, although oblique joints are also common. Some joints are open, and some have been enlarged by leaching. Both types may be coated or filled with calcite or bituminous material, or both.

FRACTURES

Fractures in the Scurry reef core differ from joints by their irregular form and more limited extent. Irregular fractures are common, and many have been enlarged by leaching. Drusy calcite and bituminous coatings are common in these fractures, and some are completely filled with calcite or bituminous material, or both.

SLUMP STRUCTURES

Contortion of limestone and claystone, resulting from slumping, has been observed in the Scurry reef both megascopically and in thin section. Megascopically the slump structures are most readily observed in core from the top part of the reef, in which black claystone emphasizes the contortion of the rock. In thin section the slump structures have been observed in calcilutite and in a few samples of unsorted calcarenite. It seems likely that slumping has occurred in many places but was not observed previously because of the monomineralic and interlensing nature of the reef limestone.

DEVELOPMENT OF REEF LIMESTONE

ACCUMULATION

The Scurry reef, like recent reefs, was a scene of intense organic activity in Permian and Carboniferous time. This reef was formed from the hard parts of organisms that lived, died, and were buried in the same local environment. A few specimens show good preservation of such delicate forms as spiny brachiopods and bryozoan fronds, indicating that at least a small amount of the unsorted débris was buried essentially *in situ*. A large part of the débris in the reef is unsorted and shows little evidence of prolonged transportation and abrasion.

The two major recognizable faunal elements in the reef are the Crinoidea and the Foraminifera, principally Fusulinidae. Corals constitute a minor element and are not abundant enough to be important here as a reef-forming organism.

The writers suggest that the welding of lime mud and the constant precipitation of calcite during reef growth rigidified the reef mass and thus substituted for the rigid organic framework that characterizes many reefs. These processes are probably adequate to form a mass rigid enough to withstand all but the most violent wave action.

Several rapidly changing reef environments existed contemporaneously in the restricted reef area, as is shown by the complex interlensing of deposits of calcilutite, angular unsorted calcarenite, rounded sorted calcarenite, oölitic limestone, and calcirudite. The calcilutite was probably deposited in quiet water. The rounded sorted calcarenite, which contains some oölites, suggests deposition in strong currents or shoaling water where wave action was effective. These turbulent conditions would account for the absence of interstitial fines and the rounding of organic débris. The angular unsorted calcarenite indicates rapid deposition of organic débris subjected to only a small amount of abrasion or winnowing. The coarse grain size, angularity, and lack of sorting of the calcirudite débris suggest break-up under violent conditions followed by relatively rapid burial.

The oölites and pisolites were probably formed on a muddy bottom that was affected by an oscillatory water motion. After they were formed by accumulation of layers of lime mud, they were either swept to a less muddy environment or were winnowed in place. In either event, the interstitial fines were removed and the final environment was one of comparatively vigorous water motion.

LITHIFICATION

Shortly after winnowing and final deposition the larger interstitial voids apparently were filled with primary crystalline cement. At this same stage or perhaps even before final deposition, some of the organic fragments and oölites apparently reorganized to form anhedral calcite mosaics. The lime mud of the calcilutite and the interstitial lime-mud filling of the unsorted calcarenite and calcirudite were probably welded by local solution and reprecipitation of carbonate particles in immediate contact, although a small amount of primary chemical cementation may have taken place in the lime mud. The stylolitic contacts of some of the fragments in the calcirudite were produced by compaction.

Cracks in semi-lithified reef sediment that were caused by settling, desiccation, and slumping were apparently filled with crystalline carbonate at the same time that the interstitial voids were being filled. Many hollow fossils, such as fusulinids, were also filled with crystalline calcite at that time and some of the cracks were filled with argillaceous material.

EARLY REPLACEMENT

The calcilutite and the interstitial filling of the calcarenite were probably dolomitized during or shortly after lithification. This dolomitization was incomplete and occurred only in scattered areas.

A small amount of silicification followed, or may have even accompanied, dolomitization. Very minute crystals of secondary feldspar may also have been formed during this stage.

INTERSTRATAL CHANGES

Fracturing followed lithification. This fracturing may have been caused by tectonic forces or by differential compaction.

Leaching was the next important process in the diagenesis of the limestone. The leaching solutions developed vugs, perhaps partly by enlargement of primary porosity; enlarged the fractures; and leached along the stylolites.

Voids, developed and enlarged by leaching, were later partly filled by calcite and rarely by dolomite. Unfilled parts of a few vugs were then filled with crystalline quartz. Additional silicification of carbonate constituents may have accompanied the quartz deposition.

The exact time at which oil entered the Scurry reef is not known. The fact that almost all of the porosity is secondary, and the absence of traces of oil stain within the secondary quartz indicate that the oil entered during a stage of development subsequent to those discussed in this paper.

REFERENCES

CLARKE, F. W., AND WHEELER, W. C. (1917), "The Inorganic Constituents of Marine Invertebrates," *U. S. Geol. Survey Prof. Paper 102*. 56 pp.

DARWIN, CHARLES, *On the Structure and Distribution of Coral Reefs*. Ward, Lock Bowden and Company, London.

HECK, W. A., YENNE, K. A., AND HENBEST, L. G. (1952), "Pennsylvanian and Permian(?) Contact in Subsurface Scurry Reef, Scurry County, Texas," *Bull. Amer. Assoc. Petrol. Geol.*, Vol. 36, pp. 1465–66.

PETTIJOHN, F. J. (1949), *Sedimentary Rocks*. 526 pp. Harper and Brothers.

ROTHROCK, H. E., BERGENBACK, R. E., MYERS, D. A., STAFFORD, P. T., AND TERRIERE, R. T., "Preliminary Report on the Geology of the Scurry Reef in Scurry County, Texas," *U. S. Geol. Survey Map OM 143*, Oil and Gas Inves. (in preparation).

TWENHOFEL, W. H. (1939), *Principles of Sedimentation*. 610 pp. McGraw-Hill Book Company, Inc.

Reprinted from
BULLETIN OF THE AMERICAN ASSOCIATION OF PETROLEUM GEOLOGISTS
VOL. 49, NO. 3 (MARCH, 1965), PP. 258-276, 7 FIGS., 18 TABLES

GEOCHEMICAL CRITERIA FOR DIFFERENTIATING REEF FROM NON-REEF FACIES IN CARBONATE ROCKS[1]

R. CHESTER[2]
Liverpool, England

ABSTRACT

The distribution of numerous trace elements has been studied in a series of reef and non-reef carbonates in order to determine their value as facies indicators. Ni, Co, Cr, V, Ba, and Sr were determined in the carbonates and all found to be significantly more abundant in the non-reef facies. These trace elements and the ratios Cr and Ni, Ni and V, and Ba and Sr may be used as facies indicators. Ni, Co, Cr, V, Pb, Ga, and Cu were determined in the separated acid-soluble fraction of the carbonates and the contents of Co and V were found to be significantly more abundant in the non-reef facies. In the acid-soluble fraction Co and V and the ratios Ni and Co, Cr and V, Pb and Ga, and Cu and Co may be used as facies indicators.

The data obtained from these carbonates of Devonian age from the Sturgeon Lake reef area of Alberta, Canada, indicate that a geochemical differentiation can be made between the carbonates of boreholes drilled in the reef facies and those of boreholes drilled in the non-reef facies.

I. Introduction

Nicholls (1958), in a general review of sedimentary geochemistry, considered the problems raised in the understanding of sedimentary rocks. These include the correlation of strata, determination of the conditions of deposition at the time the strata formed, their relative location in the basin of deposition, and that complex of environmental factors which are generally designated "facies." This paper presents the study of the variation in the trace element content of a series of sedimentary rocks of Devonian age from Alberta, Canada, in an attempt to use these trace elements as facies indicators. The rocks comprising this series are of four main types, calcareous shales, argillaceous limestones, dolomitic limestones, and dolomites, and they include two distinct facies, a reef facies and a non-reef facies. In many respects the series is ideal for a geochemical study in that the rocks comprising it have a wide lateral and a restricted vertical extent and are sampled within the same stratigraphic unit.

Graf (1960) classifies the concentration of minor elements found in carbonate rocks as a function of the amount in solid solution in the carbonate minerals and of the kinds and amounts present of detrital minerals, accessory authigenic precipitates, non-carbonate skeletal material, organic matter, phases formed during diagensis, and elements adsorbed upon all these materials. A somewhat broader genetic classification of sedimentary rocks made by Krynine (1948) may considerably simplify the classification of trace elements in such rocks as a mixture of material carried into the basin of deposition in the solid state (detrital) and material entering the basin of deposition in solution (non-detrital).

Hirst and Nicholls (1958) point out that many impure limestones can be considered to consist of two distinct fractions, an essentially carbonate fraction, such as $CaCO_3$ or $FeCO_3$, of non-detrital origin, and a detrital fraction made up essentially of silicates, clay minerals, or quartz. These two authors evolved a chemical technique for separating the detrital (acid-insoluble) fraction and the non-detrital (acid-soluble) fraction of limestones which has been used in the present study.

The concept of using geochemical criteria as indicators of the environments of deposition of sediments is not a new one. Keith and Degens (1959) investigated the difference in the trace element content of marine and non-marine shales and concluded that in the area studied it was possible to differentiate marine from fresh-water shales by quantitative spectrographic determination of B, Ga, and Rb. Potter, Shimp, and Witters (1963) determined the trace element content of 66 samples of both modern and ancient marine and fresh-water argillaceous sediments and found that B, Cr, Cu, Ga, Ni, and V are significantly more abundant in the marine than in the fresh-water sediments. Keith and Degens (1959) suggest that the chemical analysis of the total sample of a sedimentary rock is much less significant for environmental studies than analysis of separated frac-

[1] Manuscript received, July 20, 1964.

[2] Department of Oceanography, the University, Liverpool. The writer gratefully acknowledges the assistance of Imperial Oil Limited for generous financial aid and provision of samples and G. D. Nicholls of the University of Manchester, England, for suggesting the project and for helpful criticism throughout the course of the work.

tions, because interpretation requires that similar materials be compared. In the present study both the total sample (total rock) and the acid-soluble (non-detrital) fraction have been analyzed to determine the usefulness of their trace element contents as facies indicators. Included as part of this study are sedimentary rocks such as shales, the character of which can be determined by petrographic methods. These were included in order to determine the trace element content over as wide an area as possible within the basin of deposition. As the trace elements in the non-detrital portion of the rocks are removed from the overlying waters it follows that they should reflect the characteristics of the chemical environment in which the sediment was laid down. The detrital fraction, on the other hand, will reflect both the nature of the source material and those sedimentary characters which depend more on the physical variables of the area, such as current action and distance from source, and it is unlikely that the bulk composition of the detrital fraction will be changed by the environment of deposition. An exception to this is the formation of certain authigenic and degraded clay minerals.

If there is a significant variation in the content of a minor element when traced across a basin of deposition, and if this variation can be related to a certain type of environment, then that element may be useful as a geochemical indicator.

II. Geology of Sturgeon Lake Area[3]

GENERAL

The Sturgeon Lake reef lies approximately 180 miles northwest of Edmonton in northwestern Alberta between Townships 67 and 72 in the Ranges 21, 22, 23, 24, and 25 west of the Fifth Meridian. It was discovered in 1952 and contains three oil fields, designated from north to south as Sturgeon Lake, Sturgeon Lake South, and Little Smoky. The Sturgeon Lake reef is located in an embayment of the Normandville reef which lies approximately 12 miles west of it at its nearest point.

GEOLOGICAL BACKGROUND

During Late Devonian time a broad epicontinental sea covered a large expanse of western Canada. Thick sequences of limestone and shale were deposited in this area which is known as the Western Canada sedimentary basin.

Throughout much of the Upper Devonian the Alberta region was a site of flourishing reef growth. Here, the recurrent shift in the balance between the downwarping of the area and the rate of growth resulted in the alternate formation of the sheet reef, or biostrome, and the true pinnacle-type reef, the bioherm.

The best known development of the biohermal reefs is in the Edmonton area of Alberta where they grew to more than 600 feet in thickness. These bioherms are overlain and surrounded by calcareous shales containing some thin beds of carbonates.

GENERAL STRATIGRAPHY OF STURGEON LAKE AREA

A simplified stratigraphical succession is shown in Figure 1.

PRE-DEVONIAN

Quartzose and arkosic sandstones interbedded with green and maroon waxy shales, collectively referred to as the Granite Wash, underlie the Upper Devonian and overlie the Precambrian strata in this area. The thickness of the Granite Wash ranges up to 250 feet. Oil of commercial importance has been found in the sandstones.

EARLY UPPER DEVONIAN

Overlying the Granite Wash is a sequence of dark and light brown limestones with some interbedded dark gray and greenish gray calcareous shales about 300 feet thick.

UPPER DEVONIAN

The Upper Devonian sequence in the Sturgeon Lake area consists of the following stratigraphic units.

1. Woodbend Group.—The Woodbend Group contains the Ireton Formation comprising green shale and the Leduc Formation, which is the productive zone in the Sturgeon Lake, South Sturgeon Lake, and Little Smoky oil fields. The Ireton Formation has a thickness ranging from 200 to 384 feet in the productive wells. On the

[3] The description of the geology of the Sturgeon Lake area described in this section is a summary of data supplied by Imperial Oil Limited, and of the published report: Humphries, R. G., and Hage, C. O., June, 1955, "Geology, Development and Reserves, Sturgeon Lake Oil Fields, Alberta:" Canadian Mining and Metall. Bull.

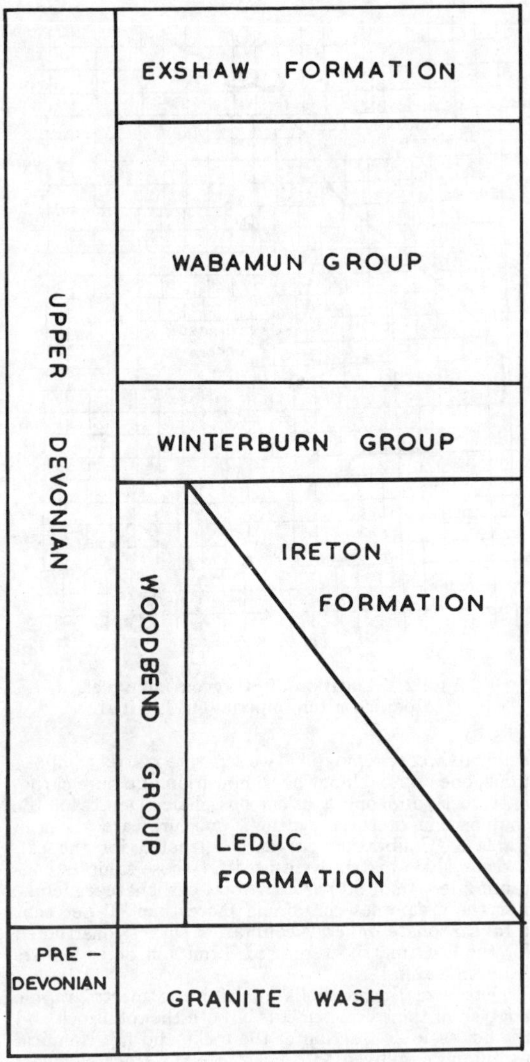

FIG. 1.—Simplified stratigraphical succession in the Sturgeon Lake area. (Modified after R. G. Humphries and C. O. Hage, 1955)

highest part of the non-productive zone of the reef to the west of Sturgeon Lake oil field, the carbonates of the reef merge with the carbonates of the overlying Winterburn Group without a separating shale. The off-reef thickness of the Ireton shale is about 1,000 feet. A dolomite and limestone unit approximately 100 feet thick occurs in the upper part of the Ireton Formation south of the Sturgeon Lake oil field. This unit in many places contains some crystalline porosity but to date there is no production from it.

The Leduc Formation is composed mainly of light tan, crystalline dolomite with some interbedded limestone and a minor amount of shale. Crystalline and vuggy porosity is fairly uniformly distributed throughout the section, although dense zones alternate with the porous zones.

The lithology of the reef within the productive oil fields is a fairly uniform dolomite section, although a great variation in the section is found in the wells located on the non-productive portions of the area. In the wells along the west side of the reef, the section is mainly dolomite but contains some limestone intervals.

2. *Winterburn Group.*—This stratigraphic unit consists of the Grammia, Calmar, and Nisku Formations. Silty dolomite and siltstone are characteristic of the Grammia and Calmar Formations, which have a combined thickness of less than 50 feet. The Nisku Formation is composed of brown crystalline dolomite, commonly silty, and has an average thickness of about 200 feet.

3. *Wabamun Group.*—This consists of light buff, fine-grained, dense, partly algal limestone and minor dolomite and has an average thickness of 800 feet.

4. *Exshaw Formation.*—This consists of dark brown to black bituminous and calcareous shale and has an average thickness of about 20 feet.

MISSISSIPPIAN

The Mississippian succession, which ranges in thickness from 1,500 to 2,000 feet, is divided into an upper Rundle Group and a lower Banff Formation. The Rundle Group is composed of limestone, dolomite, shale, and small amounts of anhydrite. The Banff Formation, which ranges in thickness from 650 to 750 feet, is composed mainly of argillaceous limestone with a calcareous unit at the top.

GEOLOGICAL HISTORY OF STURGEON LAKE REEF

The first stage in the history of the reef is represented by a series of argillaceous limestones and calcareous shales which were deposited in fairly shallow water. This was followed by a steady sinking of the sea floor during which reef growth was initiated. The result of this was the formation of a typical transgressive reef—the Lower Sturgeon reef,—which has an average thickness of approximately 800 feet. Although the reef is mainly dolomite, non-reef limestone has been found in two localities toward the base of the reef section. Gray and green shales ranging from 650 to 900 feet in thickness and containing minor fragmental limestones are found around and partly overlying the Lower Sturgeon reef. The lower half of these shales is usually markedly calcareous and is considered to be contemporaneous with the reef growth.

The next phase in the history of the reef involved a tilt toward the northeast which resulted in the cessation of growth along the eastern rim of the reef although the western rim continued to grow. Normal marine sediments were deposited around and over the submerged rim although a fair amount of reef detritus was laid down closer to the living reef, which is named the Upper Sturgeon reef. At this stage the sea was shallow and the Upper Sturgeon reef shows some of the characteristics of a regressive reef. Toward the northeast the Upper Sturgeon reef gives way to off-reef limestones which gradually become more argillaceous with intercalations of shale. These are overlain by a banded sequence of non-reef argillaceous limestones and calcareous shales. The final shallowing of the Upper Devonian before the deposition of the limestones of the Wabamun Group is represented in the deposition of dolomite with some limestones and frequent intercalations of siltstone or fine sandstone. By this time all normal reef growth had stopped, the sea was very shallow and occasional desiccation took place.

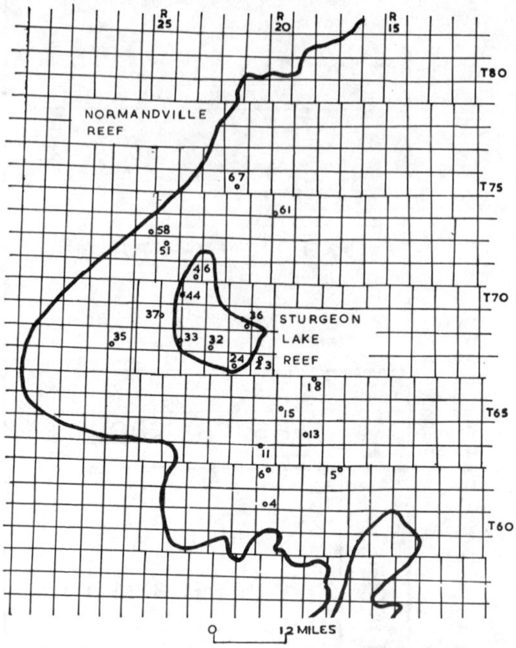

FIG. 2.—Location of Sturgeon Lake reef. (Modified after Imperial Oil Limited)

III. METHODS OF ANALYSIS

1. Selection of samples.—The rock samples used in the present study are from the Ireton and Leduc Formations of the Woodbend Group. They are selected from the Lower Sturgeon reef, or from strata equivalent to it in the non-reef areas, within the interval extending from the base of the reef, or its equivalent in the non-reef areas, up to but excluding post-reef samples. The boreholes selected are illustrated in Figure 2 and their map references listed in Appendix III.

2. Preparation of samples.—Each borehole sample, which consisted of small drill cuttings ranging between $\frac{1}{8}$ and $\frac{1}{2}$ inch in diameter, represents 10 or 20 feet of rock. The samples were washed in distilled water, dried and crushed to pass a 90-mesh BSS sieve.

3. Trace element determination.—Trace element determinations were made using a Hilger E. 492 fully automatic large-quartz spectrograph fitted with a Hilger F. 522 fully enclosed general purpose arc and spark stand. For the purpose of emulsion calibration a rotating stepped sector was placed between the lens and the slit. A Joyce-Loebl recording microdensitometer was used for the measurement of spectral line intensities.

Trace element determinations on the total rock samples were made after heating the rock powders to 900°C. to drive off CO_2 and to convert them to a mixture of oxides in an attempt to have all samples in a common matrix. Trace element contents were recalculated for the loss of weight on heating. Standards for the compilation of calibration curves for the total rock trace elements were prepared in two separate matrix compositions, one approaching the composition of a pure carbonate rock, and one a calcareous shale. Details of the composition of these synthetic mixtures are given in Table I. Calibration curves constructed for the calcareous shale matrix were used for those samples containing less than 50 per cent total carbonate minerals. For those samples containing more than 50 per cent total carbonate minerals calibration curves constructed for the CaO matrix were used. Limits of detection are listed in Table II.

Details of the methods used for the spectrographic analysis of the trace elements of both the total rock and the acid-soluble fraction of the rocks and the chemical technique employed to concentrate the trace elements of the acid-soluble fraction were essentially similar to those used by Hirst and Nicholls (1958) and have been fully described. The chemical procedure utilizes an

TABLE I. MIXTURES USED FOR PREPARATION OF SYNTHETIC STANDARDS

	Calcareous Shale Matrix wt. %	*Carbonate matrix wt. %*
CaO	36.0	100
Al_2O_3	15.0	—
SiO_2	42.0	—
TiO_2	0.7	—
Total iron	2.5	—
MgO	1.0	—
Na_2CO_3	0.4	—
K_2CO_3	2.4	—
	100.0	100

TABLE II. LIMITS OF DETECTION IN TRACE ELEMENT DETERMINATIONS

Element	Wave-length Å	Detection limit
Ni	3,414.765	Detection to 6 ppm Fe_2O_3 matrix, 6 ppm Al_2O_3 matrix, 5 ppm CaO matrix, 10 ppm shale matrix.
Co	3,453.505	Detection to 6 ppm Fe_2O_3 matrix, 6 ppm Al_2O_3 matrix, 5 ppm CaO matrix, 10 ppm shale matrix.
V	3,185.396 4,379.238	Detection to 20 ppm Fe_2O_3 matrix (line 3,185.396), 10 ppm Al_2O_3 matrix (line 4,379.238), 6 ppm CaO matrix (line 4,379.238), 10 ppm shale matrix (line 4,379.238).
Cr	4,254.346	Detection to 6 ppm Fe_2O_3 matrix, 3 ppm Al_2O_3 matrix, 6 ppm CaO matrix, 10 ppm shale matrix.
Pb	4,057.820	Detection to 30 ppm Fe_2O_3 matrix, 30 ppm Al_2O_3 matrix.
Ga	4,172.056	Detection to 30 ppm Fe_2O_3 matrix, 20 ppm Al_2O_3 matrix.
Cu	3,247.540	Detection to 10 ppm Fe_2O_3 matrix, 10 ppm Al_2O_3 matrix.
Ba	4,934.086	Detection to 20 ppm CaO matrix, 30 ppm shale matrix.
Sr	4,607.331	Detection to 15 ppm CaO matrix, 20 ppm shale matrix.
Pd	3,481.152	Internal standard used for the elements Co, Ni, Cr, V, Ba, and Sr.
In	4,511.323	Internal standard used for the elements Cu, Pb, and Ga.

initial attack by 25 per cent v/v acetic acid and extensive checks were made on the residues obtained from this procedure to determine if solution of the carbonate mineral was complete. Although it was found that solution was incomplete for certain samples having a high proportion of dolomite it was not thought desirable to increase the strength of the acid because of the possibility of attacking the clay mineral lattices.

Samples from boreholes number 24, 36, and 33 showed traces of dolomite in the acid-insoluble residues and the carbonate content of these samples was determined with 2N HCl. The results are listed in Table III. The non-detrital (acid-soluble) fraction liberated from those samples which show incomplete solution of the carbonate minerals will not be the complete non-detrital fraction. However, in a rock having high carbonate content, which consists predominantly of one carbonate mineral, the trace element content has been assumed to be evenly distributed through the carbonate fraction and the proportion dissolved in the initial acid attack representative of this fraction to a first approximation. The non-detrital trace element analyses of those samples showing incomplete solution of the carbonate fraction were rejected unless the carbonate fraction made up more than 90 per cent of the sample and consisted of one mineral only, i.e., either calcite or dolomite. Those samples which were rejected are marked with an asterisk in Table III.

Calculation of the absolute trace element contents of the soluble fraction of the rocks, from the determinations on the ashed concentrates, is made by applying the formula

$$\left\{\begin{matrix}\text{Content of element E} \\ \text{in the soluble fraction}\end{matrix}\right. = N\frac{X}{x}\cdot\frac{y}{Y}$$

where the ashed concentrate (diluted with Al_2O_3 or Fe_2O_3 to give, when the amount of ashed concentrate permitted, a volume sufficient to fill two electrodes) contains N ppm of an element E; x gm. is the weight of ashed concentrate added to X−x gm. of diluent material to give X gm. of analysis material; y gm. is the total weight of ashed concentrate obtained from Y gm. of soluble fraction, where Y is the difference in weight of original rock taken and the weight of residue left after the acetic acid attack. The ashed concentrate obtained in this method was split into two fractions, one for the determination of the involatile elements, viz., Co, Ni, V, and Cr and one for the determination of the volatile elements, viz., Cu, Pb, and Ga.

In certain cases the trace elements were present in amounts below the limit of detection of the spectrographic method employed. Borehole averages obtained by taking no account of these determinations could yield anomolous results, as could averages obtained by assigning such results a value of zero ppm. In such cases, therefore, the trace element content was estimated by assigning the sample half the value between zero ppm and the limit of detection for a particular element. This figure was then used in the appropriate calculation, i.e., in correcting for the total rock loss on heating or designated N is the calculation for the acid-soluble fraction. The trace element values which were estimated in this manner are listed in Appendices I and II.

4. X-ray diffraction analysis.—A composite representative sample of approximately 0.2 gm. was obtained for each borehole by a simple mixing and quartering technique. From this the average mineral composition of the rocks in the borehole was obtained. The rock powders had previously passed a 90-mesh BSS sieve and were further pulverized in an agate mortar. All determinations were made in the form of smears mounted in acetone on glass slides and no attempt was made to control the amount of power on each slide. The determinations were made using a Phillips diffractometer with the following instrument settings: ratemeter—4, slits—1°, chart speed—800, rate of scanning—$\frac{1}{2}$° per minute, radiation—$CuK\alpha$.

Weber and Smith (1961) have determined calcite-dolomite ratios in carbonate rocks by comparing the areas of the strongest diffraction peaks of calcite and dolomite and this principle was adopted by the present writer. A series of standards of various mixtures of natural calcites and dolomites was prepared, the intensities of the diffraction maxima of these mixtures

TABLE III. DETERMINATION OF CARBONATE CONTENT OF SAMPLES CONTAINING DOLOMITE IN ACID-INSOLUBLE RESIDUES

Sample	% Soluble in 25% Acetic Acid	Nature of Residue from Acetic Acid Attack	% Soluble in 2N HCl
Borehole No. 24			
9,250	37.9	Dolomite+quartz+illite	92.9
9,330	49.5	Dolomite+quartz	98.6
9,410	36.0	Dolomite+quartz+illite	91.8
9,490	37.0	Dolomite+quartz	97.1
Borehole No. 36			
8,510	72.0	Quartz+illite	—
8,580	44.2	Dolomite+quartz+illite	97.9
8,660	47.4	Dolomite+quartz	98.2
Borehole No. 33			
8,240*	63.0	Dolomite+quartz+illite	88.2
9,320	65.0	Dolomite+quartz+illite	97.9
9,400*	47.0	Dolomite+quartz+illite	77.0
9,480	64.0	Dolomite+quartz+illite	85.3
9,560	54.7	Dolomite+quartz+illite	94.5
9,640*	53.1	Dolomite+quartz+illite	74.8
9,720	57.6	Dolomite+quartz+illite	95.0
9,810	73.0	Dolomite+quartz+illite	91.8
9,970*	39.2	Dolomite+quartz+illite	70.3
10,050	57.2	Dolomite+quartz+illite	95.2
10,130	54.0	Dolomite+quartz	98.8
10,210*	35.1	Dolomite+quartz+illite	72.0
10,290	38.5	Dolomite+quartz+illite	95.9

* Samples rejected.

TABLE IV. MINERALOGY OF ROCKS

Borehole	*Per Cent Total Carbonate*	*Dolomite as Percentage of Total Carbonate*	*Illite*	*Chlorite*	*Illite-Chlorite Ratio*	*Type of Sedimentary Rock*
24	95	98	√	×	—	Dolomite
11	43	16	√	×	—	Calcareous dolomitic shale
6	42	20	√	×	—	Calcareous dolomitic shale
15	38	0	√	√	0.63	Calcareous shale
4	54	0	√	√	2.0	Argillaceous limestone
13	55	0	√	√	3.0	Calcareous shale
12	42	0	√	√	1.3	Calcareous shale
5	41	11	√	√	1.4	Calcareous dolomitic shale
33	87	94	√	×	—	Dolomite
37	72	15	√	√	1.6	Dolomitic limestone
35	51	8	√	√	1.7	Argillaceous limestone
44	87	42	√	×	—	Dolomitic limestone
58	51	9	√	√	1.5	Argillaceous limestone
51	44	17	√	√	3.5	Calcareous dolomitic shale
46	83	49	√	×	—	Dolomitic limestone
67	28	0	√	√	1.6	Calcareous shale
61	32	11	√	√	1.5	Calcareous dolomitic shale
32	83	33	√	×	—	Dolomitic limestone
36	89	90	√	×	—	Dolomite
23	36	13	√	√	2.8	Dolomitic shale
18	29	30	√	√	1.2	Calcareous dolomitic shale

√ = clay mineral present.
× = clay mineral absent.

measured (as peak areas), and a calibration curve was drawn up showing the relation between the weight of calcite and dolomite and their relative diffraction intensities. The strongest X-ray diffraction spectra of calcite and dolomite, at 29.4° 2θ, and 30.96° 2θ, respectively, were measured. The calibration curve was drawn after the method of Weber and Smith (1961).

Illite, chlorite, and quartz were identified by their strongest diffraction peaks and clay mineral ratios were obtained by direct comparison of peak areas.

IV. MINERALOGY OF SEDIMENTS

The boreholes studied in the Sturgeon Lake reef area may be divided into two general groups: (1) the Sturgeon Lake reef facies, which includes those boreholes located on the Sturgeon Lake reef, viz., Nos. 24, 33, 44, 46, 32, and 36; and (2) the non-reef facies, which includes all those boreholes not located on a reef area, viz., Nos. 11, 6, 15, 13, 12, 35, 51, 67, 61, 23, and 18.

The rocks of this area are composed almost entirely of five minerals: calcite, dolomite, quartz, and the clay minerals illite and chlorite.

For the purposes of this paper the following nomenclature is used in sedimentary rock classification.

The term *dolomite* is used to designate a rock having a total carbonate content of more than 70 per cent, of which 90 per cent or more consists of dolomite.

The term *dolomitic limestone* is used to designate a rock having a total carbonate content of more than 70 per cent, of which 10–15 per cent is dolomite.

The term *limestone* is used to designate a rock having a total carbonate content of more than 70 per cent, of which more than 90 per cent is calcite.

The term *argillaceous limestone* (or *dolomite*) designates a rock having a non-carbonate fraction (containing quartz and one or more of the clay minerals) of 30–50 per cent.

The term *shale* (calcareous or dolomitic) is used to designate a rock having a non-carbonate content of 50–80 per cent which consists of quartz and one or more of the clay minerals. The rocks of various boreholes are designated according to this scheme in Table IV.

Calcite and dolomite are the only carbonate minerals found in any of the rocks investigated. The total carbonate content, estimated by determining the percentage of the rock soluble in 25 per cent acetic acid (except for those samples containing a high proportion of dolomite where hydrochloric acid was also used) ranges between 28 and 99 per cent. The average total carbonate content of each borehole is listed in Table IV.

The dolomite content of the total carbonate minerals ranges between zero per cent in certain of the non-reef rocks to a maximum of 98 per cent in the rocks of borehole number 24 on the Sturgeon

TABLE XI. AVERAGE NON-DETRITAL (ACID-SOLUBLE) TRACE ELEMENT CONTENT OF BOREHOLE ROCK SECTIONS

Borehole	*Ni*	*Co*	*Cr*	*V*	*Cu*	*Pb*	*Ga*	*Ni/Co*	*Cr/V*	*Pb/Ga*	*Cu/Co*
24	23	0.38	3.6	1.7	18	29	16	61	2.1	1.8	47
11	82	4.2	6.1	11	26	34	20	20	0.55	1.7	6.2
6	45	4.4	4.9	8.3	18	30	16	10	0.59	1.9	4.1
15	30	5.1	5.8	7.0	—	—	—	5.9	0.83	—	—
13	19	3.7	3.8	5.8	15	17	7.0	5.1	0.66	2.4	4.1
12	16	3.4	4.0	4.8	14	19	13	4.7	0.83	1.5	4.1
5	17	3.5	2.6	3.8	—	—	—	4.9	0.68	—	—
33	13	0.49	3.4	3.0	24	8.6	18	27	1.1	0.48	49
37	15	1.6	3.0	3.3	—	—	9.9	9.4	0.91	—	—
35	18	2.9	2.7	8.1	16	9.4	—	6.2	0.33	0.95	5.5
44	14	0.57	1.4	3.2	—	—	—	25	0.44	—	—
58	—	—	—	—	23	23	12	—	—	1.9	—
51	47	3.2	7.8	9.6	15	21	12	15	0.81	1.7	4.7
46	8.0	0.92	0.88	2.9	18	11	4.7	8.7	0.30	2.3	20
67	27	7.6	6.5	10	20	19	17	3.6	0.65	1.1	2.6
61	31	8.1	6.4	12	27	32	15	3.8	0.53	2.1	3.3
32	51	1.1	6.6	4.0	19	4.7	9.8	46	1.6	0.48	17
36	22	0.06	2.7	1.2	—	—	—	367	2.2	—	—
23	59	5.2	9.1	8.4	—	—	—	11	1.1	—	—
18	32	8.6	8.1	13	23	17	11	3.7	0.62	1.5	2.7

Certain trace element values used in calculating borehole averages were below the limit of detection of the spectrographic method and were therefore estimated. (See text and Appendix II)

the distribution of the trace elements of the non-detrital fraction are given in Figure 6, and the partition values selected by inspection of these are listed in Table XIII.

The results of the Student's t-tests are listed in Table XIV, from which it would appear that Co and V are possible facies indicators. The usefulness of Pb as a facies indicator was also considered as it shows a tendency to be concentrated in the non-reef facies even though the difference between the two facies is not significant at the 1 per cent level.

Figure 4 illustrates the variations of the trace elements present in the non-detrital fraction between the reef and the non-reef facies.

Co and V, which are Class II elements, exhibit

TABLE XII. AVERAGE NON-DETRITAL (ACID-SOLUBLE) TRACE ELEMENT CONTENT OF SEDIMENTARY ROCKS AND FACIES

I. The Sediments

	Ni	*Co*	*Cr*	*V*	*Cu*	*Pb*	*Ga*	*Ni/Co*	*Cr/V*	*Pb/Ga*	*Cu/Co*
Dolomites	19	0.31	3.2	2.0	21	19	17	61	1.6	1.1	6.8
Dolomitic limestones	22	1.0	3.0	3.3	18	7.8	7.3	22	0.91	1.1	18
Argillaceous limestones	18	3.3	3.2	6.9	18	16	9.6	5.5	0.46	1.7	5.5
Shales	39	6.1	6.1	8.8	20	25	15	7.4	0.69	1.7	3.8

II. The Facies

	Reef Facies		*Non-Reef Facies*	
Ni	22	(0.41–100)	34	(4.9–140)
Co	0.59	(0.03–4.4)	4.7	(0.18–18)
Cr	3.1	(0.21–29)	5.4	(0.11–16)
V	2.7	(0.30–18)	8.1	(0.60–36)
Cu	20	(2.6–50)	20	(4.0–88)
Pb	13	(0.58–68)	22	(1.1–96)
Ga	12	(0.50–54)	13	(0.35–55)
Ni/Co	37	(2.7–544)	7.2	(0.66–209)
Cr/V	1.1	(0.12–17)	0.67	(0.01–4.8)
Pb/Ga	1.1	(0.30–3.5)	1.7	(0.33–88)
Cu/Co	34	(1.2–136)	4.3	(0.45–37)

Figures in parentheses refer to the minimum and maximum values in each facies.

certain similarities in their variation among the rocks, and when they are considered as element pairs with the elements of Class I, it should be possible to use the ratio of each pair as a facies indicator. Similarly the values of the Pb and Ga element pair are considered as facies indicators. The ideal facies indicator ratio would be an element pair consisting of one element which is concentrated in the non-reef facies and one concentrated in the reef facies. In the absence of an element which is concentrated in the reef facies it is possible instead to use an element which exhibits a minimum amount of variation between the two facies. Gallium was such an element but its variation among the rock types complicated its over-all distribution. Copper, on the other hand, exhibits the least amount of variation either among the sediments or between the two facies, and this element should provide a standard against which the variation of an element concentrated in a particular facies can be measured. Cobalt was selected as the best example of an element concentrated in the non-reef facies and the Cu:Co ratio was examined as a possible facies indicator.

Table XV gives the percentage of samples which is correctly classified by use of the partition values assigned to the trace elements and trace element ratios.

The percentages of the samples correctly classi-

TABLE XIII. NON-DETRITAL FRACTION PARTITION VALUES (in ppm)

	Reef Facies	*Non-Reef Facies*
Co	< 2	> 2
V	< 4	> 4
Pb	<10	>10
Ni/Co	>20	<20
Cr/V	> 1	< 1
Pb/Ga	< 1	> 1
Cu/Co	>15	<15

TABLE XIV. NON-DETRITAL FRACTION STUDENT'S t-TESTS

Type of Test		*t* .995	*t Calculated*	*Significant at 1% Level*
V	modified t	2.67	8.3	Yes
Cr	modified t	2.62	0.7	No
Ni	modified t	2.62	1.0	No
Co	modified t	2.63	11.6	Yes
Pb	t test	2.62	1.7	No
Ga		Means equal		No
Cu	modified t	2.71	0.3	No

TABLE XV. PERCENTAGE OF REEF AND NON-REEF SAMPLES CORRECTLY CLASSIFIED USING NON-DETRITAL FRACTION PARTITION VALUES

	Samples from Reef Facies	*Samples from Non-Reef Facies*	*Samples from Both Facies*
Co	93	88	89
V	79	75	76
Pb	52	70	66
Ni/Co	76	90	85
Cr/V	55	76	70
Pb/Ga	44	77	70
Cu/Co	74	97	91

TABLE XVI. PERCENTAGE OF BOREHOLE ROCK SECTIONS CORRECTLY CLASSIFIED, USING NON-DETRITAL FRACTION PARTITION VALUES

	Borehole Rock Sections from Reef Facies	*Borehole Rock Sections from Non-Reef Facies*	*All Borehole Rock Sections*
Co	100	92	95
V	100	85	89
Pb	50	90	79
Ni/Co	83	92	89
Cr/V	67	92	84
Pb/Ga	50	90	79
Cr/Co	100	100	100

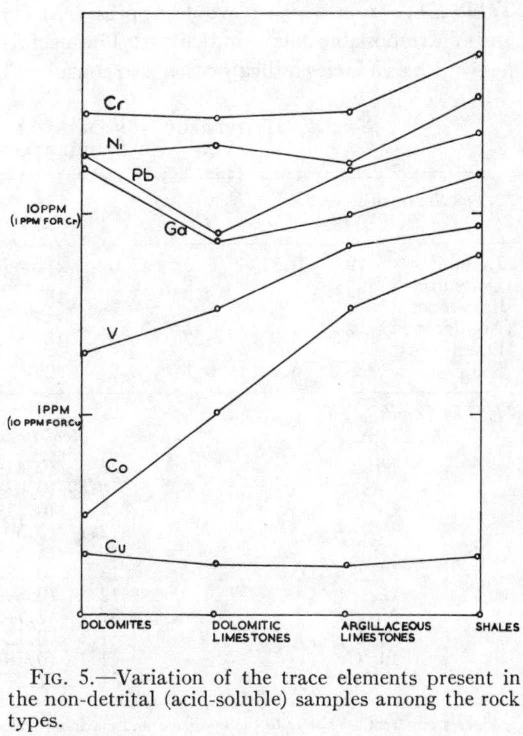

FIG. 5.—Variation of the trace elements present in the non-detrital (acid-soluble) samples among the rock types.

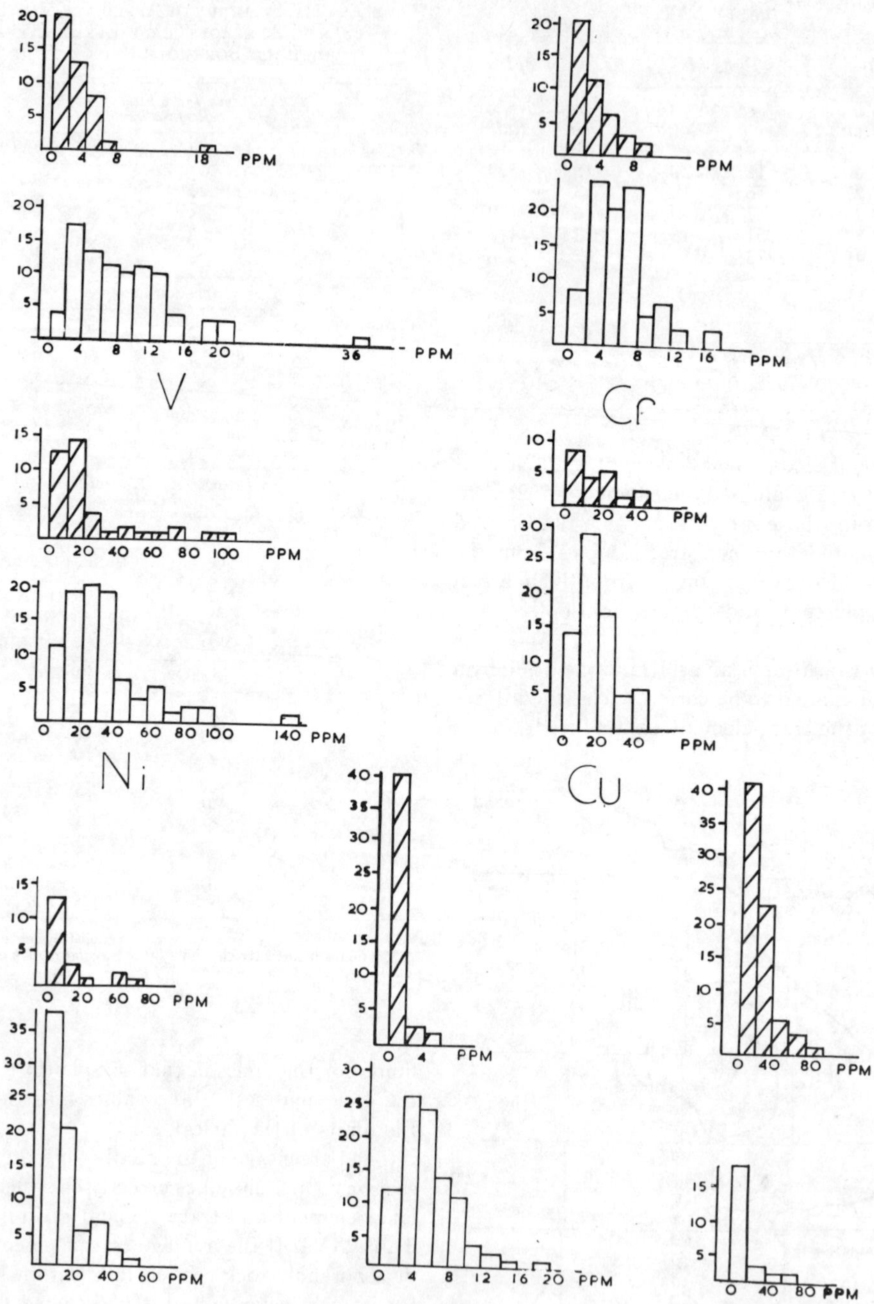

FIG. 6.—Histograms illustrating the distribution of the trace elements of the non-detrital (acid-soluble) samples between the reef and non-reef facies.

TABLE XVII

Type I		Type II		Type III	
Bore-hole	*Ni/Co Ratio*	*Bore-hole*	*Ni/Co Ratio*	*Bore-hole*	*Ni/Co Ratio*
24	60	11	20	15	5.9
33	27	6	10	13	5.1
44	25	37	9.4	6	4.9
46	87	51	15	12	4.7
32	46	23	11	67	3.6
				61	3.8
				18	3.7
				35	6.2
		Characteristic Ratio			
>25		10–20		<10	

fied by use of a combined 8-element (and element ratio) test or a combined 4 element (and element ratio) test for those samples in which Pb, Ga, and Cu were not determined are: samples from the reef facies, 71 per cent; samples from the non-reef facies, 87 per cent; and samples from both facies, 82 per cent.

This classification is an arbitrary one whereby a sample is assumed to be correctly classified if five or more of the trace element values lie within the

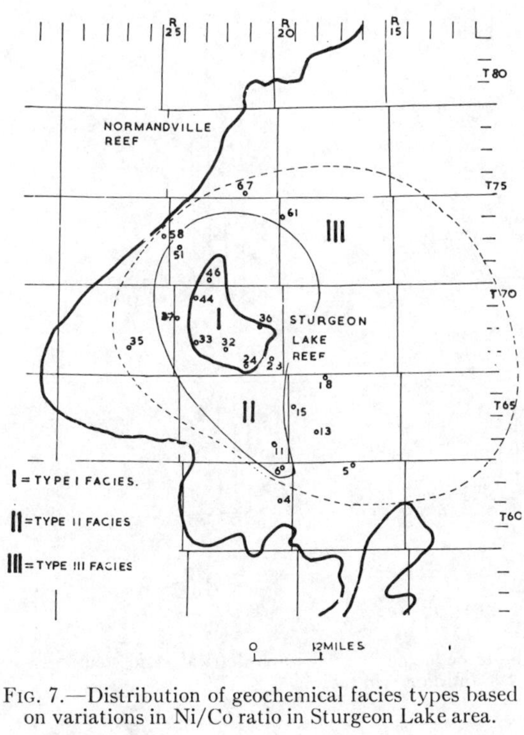

FIG. 7.—Distribution of geochemical facies types based on variations in Ni/Co ratio in Sturgeon Lake area.

TABLE XVIII. SUMMARY OF APPLICATION OF PARTITION VALUES TO AVERAGE TRACE ELEMENT CONTENT OF EACH BOREHOLE ROCK SECTION

A. Total Rock Trace Element Indicators

Borehole	*Ni*	*Co*	*V*	*Cr*	*Ba*	*Sr*	*Cr/Ni*	*Ni/V*	*Ba/Sr*
Reef Facies									
24	√	√	√	√	√	√	√	√	√
33	√	√	√	√	√	√	√	√	√
44	√	×	√	√	√	√	√	√	×
46	√	√	√	√	√	√	√	√	×
32	×	√	√	√	√	×	√	√	×
36	√	√	√	√	√	√	√	√	√
Non-Reef Facies									
11	√	√	√	√	√	√	√	√	√
4	√	√	√	√	√	√	√	√	√
13	√	√	√	√	√	√	√	√	√
19	√	√	√	√	√	√	√	√	√
37	×	√	√	√	√	√	√	√	√
35	√	√	√	√	√	√	√	√	√
51	√	×	√	√	√	√	√	√	√
67	√	√	√	√	√	√	√	√	√
18	√	√	√	√	√	√	√	√	√

B. Non-Detrital Trace Element Indicators

Borehole	*Co*	*V*	*Pb*	*Ni/Co*	*Cr/V*	*Pb/Ga*	*Cu/Co*
Reef Facies							
24	√	√	×	√	√	×	√
33	√	√	√	√	√	√	√
44	√	√	N.D.	√	×	N.D.	N.D.
46	√	√	×	×	×	×	√
32	√	√	√	√	√	√	√
36	√	√	N.D.	√	√	N.D.	N.D.
Non-Reef Facies							
11	√	√	√	×	√	√	√
6	√	√	√	√	√	√	√
15	√	√	N.D.	√	√	N.D.	N.D.
13	√	√	√	√	√	√	√
5	√	×	√	√	√	√	√
12	√	√	√	√	√	√	√
37	×	×	N.D.	√	√	N.D.	N.D.
35	√	√	×	√	√	×	√
51	√	√	√	√	√	√	√
67	√	√	√	√	√	√	√
61	√	√	√	√	√	√	√
23	√	√	N.D.	√	×	N.D.	N.D.
18	√	√	√	√	√	√	√

√ = facies correctly classified by the average trace element content.
× = facies incorrectly classified by the average trace element content.
N.D. = no data.

limits for the facies of that sample (in the case of the 4-element test, three values must fall within the limits of the facies).

The percentage of the number of borehole rock sections which may be correctly classified by each trace element and trace element ratio is listed in Table XVI. If the average trace element content of a borehole rock section falls within the limits for its particular facies, it is assumed to be correctly classified.

The Ni/Co ratios of the various boreholes may be divided into three general groups (Table XVII).

Type I consists of those rocks which belong to the reef facies. Type III consists of those rocks

which belong to the non-reef facies and Type II consists of those non-reef rocks which are transitional between those of the reef facies and those of Type III.

Ni is concentrated in the non-reef facies but not to the same extent as Co or V. Within these facies, however, the distribution of Ni can be divided into two general groups, showing a rise in the Ni content in those rocks which are, in general, closer to the reef. In the reef area, however, the Ni content again falls. In contrast to Ni, Co is strongly concentrated in the non-reef facies and within it has a fairly even distribution. The Ni/Co ratio is therefore a good indicator of the transi tional rocks.

In the non-reef facies Types II and III are independent of sedimentary rock types, but of the trace elements investigated Ni is the only one to show this relationship. In the non-reef facies the distribution of the other trace elements in the non-detrital fraction of the samples is dependent on the rock type in which they occur.

Because the Ni/Co ratios allow the rocks of the non-reef facies to be divided into two types, it is possible that a "transitional" facies may be introduced within the non-reef facies. It must be stressed, however, that this division is a geochemical one and that only the distribution of Ni and the Ni/Co ratio are diagnostic of it, and that other trace element parameters must be discovered before such a geochemical facies can be fully established. The division of the Sturgeon Lake reef area into the three types is illustrated in Figure 7.

VIII. Conclusions

Ingerson (1962) has noted that a current problem in the geochemistry of sedimentary carbonate rocks is the characterization of reef-related facies, in such a way that data from one well penetrating a reef assemblage can be used to discover in which facies—backreef, reef, lagoonal, basin—the well is situated. Data obtained from the Sturgeon Lake area indicates that such a geochemical differentiation is possible between rocks from boreholes drilled in the reef facies and those drilled in the non-reef facies.

Table XVIII summarizes the geochemical data for both the total rock trace elements and those present in the non-detrital fraction.

Appendices I and II

Trace Element Analysis

Values given as parts per million.
N.D. = Not detected.
* = Determinations made on a single arcing. In Appendix II this refers to the determination of the elements Ga, Cu, and Pb only.
E = Trace element content has been estimated because the analysis is below the level of detection.
— = No data.
In both appendices analyses are expressed in parts per million.

Appendix I

Total Rock Trace Element Analyses

I. Reef Samples

	Borehole (*Depth in Feet*)	*Ni*	*Co*	*Cr*	*V*	*Ba*	*Sr*	*Cr/Ni*	*Ni/V*	*Ba/Sr*
24.	9,090	18	2.2 E	2.0	7.4	48	14	0.11	2.4	3.4
	9,170	25	2.7	0.7 E	2.2 E	9.6	7.4	0.03	11	1.3
	9,250	7.9	3.0 E	1.0 E	3.0 E	31	4.3 E	0.13	2.6	7.2
	9,330	3.9	2.3 E	3.2	2.0	51	23	0.82	2.0	2.2
	9,410	10	2.3 E	0.7 E	4.6	47	21	0.08	2.2	2.2
	9,490	5.1	2.4	1.7	1.8	31	12	0.33	2.8	2.6
	9,570	7.8	8.6	0.7 E	2.2 E	69	19	0.09	3.5	3.6
	9,650*	24	5.7	0.7 E	3.2 E	24	7.3	0.03	7.5	3.3
	9,730	14	5.0	0.7 E	3.4	53	7.2	0.05	4.1	7.4
	9,810*	5.6	4.6	3.7	3.9	107	39	0.66	1.4	2.7
33.	9,480	66	2.4 E	0.8 E	12	32	11	0.01	5.5	2.9
	9,560	14	2.4 E	4.7	2.4E	53	63	0.34	5.8	0.84
	9,640	13	2.4 E	0.8 E	2.4E	51	6.9 E	0.06	5.4	7.4
	9,810	2.8	2.3 E	0.9 E	3.9	70	54	0.32	0.71	1.3
	9,970	7.7	2.5 E	0.8 E	2.5E	33	21	0.10	3.1	1.5
	10,050	7.6	2.5 E	0.8 E	2.5	47	16	0.11	3.0	2.9
	10,130	15	2.4 E	4.1	2.5	40	17	0.27	6.0	2.3
	10,210	2.4	2.5 E	0.8 E	5.7	60	42	0.35	0.42	1.4
	10,290	1.4	2.5 E	0.8 E	2.5 E	50	27	0.59	0.56	1.8

Appendix I—(*continued*)

Borehole	(Depth in Feet)	*Ni*	*Co*	*Cr*	*V*	*Ba*	*Sr*	*Cr/Ni*	*Ni/V*	*Ba/Sr*
44.	8,700	13	14	7.0	5.2	49	108	0.54	2.5	0.45
	8,780	15	47	7.4	2.4	137	66	0.50	6.2	2.1
	8,840	21	15	0.9 E	2.2 E	28	56	0.04	9.5	0.49
	8,920	12	6.3	2.5	2.3 E	40	55	0.20	5.2	0.73
	9,000	6.1	6.1	2.2	8.4	43	89	0.36	0.72	0.48
	9,080	13	23	0.7 E	2.4 E	22	5.8 E	0.06	5.4	3.8
	9,160	6.2	N.D.	9.7	12	22	94	1.6	0.52	0.23
	9,240	16	N.D.	0.7 E	1.8	17	5.9 E	0.05	8.9	2.9
	9,320	9.1	19	1.1	2.3 E	79	23	0.12	4.0	3.5
46.	8,700	15	4.9	5.2	2.5 E	44	5.9 E	3.5	6.0	7.5
	8,750*	27	7.7	3.1	2.6 E	19	2.4 E	0.11	10	8.0
	8,860	2.0 E	2.2 E	4.8	2.0	33	80	2.4	1.0	0.41
	8,940	12	2.3 E	2.6	2.3 E	26	68	0.22	5.2	0.38
	9,020*	15	2.3 E	0.7 E	2.3 E	24	20	0.05	6.6	1.2
	9,100	12	2.2 E	1.4	12	24	20	0.12	1.0	1.2
	9,150	23	2.3 E	3.1	2.3 E	14	5.1 E	0.14	10	2.7
	9,260	7.5	2.2 E	0.7 E	2.2 E	15	7.5	0.10	3.4	2.0
	9,340	28	2.1 E	3.3	2.1 E	15	46	0.12	13	0.32
	9,400	22	3.3	1.5	3.7	29	22	0.07	6.0	1.3
32.	9,165	169	1.9 E	35	8.5	66	1,000	0.21	20	0.07
	9,245	8.0	2.1 E	49	2.1 E	135	1,000	6.1	3.8	0.13
	9,325	240	2.1 E	0.7 E	2.1 E	36	187	0.003	114	0.19
	9,405	9.9	2.4 E	0.8 E	2.4 E	32	21	0.08	4.1	1.5
	9,485	86	2.2 E	4.4	2.2 E	32	32	0.05	39	1.0
	9,565	118	2.1 E	18	2.1 E	33	82	0.15	56	0.40
	9,645	105	2.1 E	9.0	2.1 E	36	99	0.09	50	0.36
	9,725	116	2.2 E	14	2.2 E	40	260	0.12	5.3	0.15
	9,805	1.9	2.2 E	5.3	2.7	22	147	2.8	0.70	0.15
	9,885	39	2.2 E	23	2.1 E	38	90	0.59	19	0.42
	9,965	6.1	2.3 E	15	3.4	49	222	2.5	1.8	0.22
	10,045	5.1	2.1 E	15	15	67	—	2.9	0.34	—
36.	8,510	28	7.8	42	2.2 E	75	31	1.5	13	2.4
	8,580	19	5.7	2.9	7.0	39	4.3	0.15	2.7	9.1
	8,660	15	2.1 E	2.4	2.1 E	33	5.9	0.16	7.1	5.6
	8,740	17	5.2	2.9	2.4	38	6.7	0.17	7.1	5.7
	8,820	20	3.1	3.0	4.9	43	13	0.15	4.1	3.3
II. Non-Reef Samples										
11.	9,810	95	13	258	61	250	217	2.7	1.5	1.1
	9,890	149	11	176	70	203	477	1.2	2.1	0.43
	9,970*	104	7.8	115	85	303	999	1.1	1.2	0.30
	10,130	126	11	151	35	187	—	1.2	3.6	—
	10,290	73	5.5	158	38	141	133	2.2	1.9	1.1
4.	10,830	38	21	166	51	249	285	4.4	0.75	0.87
	10,910	35	14	37	33	209	432	1.1	1.1	0.48
	10,990	23	N.D.	56	22	156	483	2.4	1.0	0.32
	11,070	47	13	53	35	338	488	1.1	1.3	0.69
13.	8,610	21	28	27	14	95	474	1.3	1.5	0.20
	8,690*	53	18	113	58	129	173	2.1	0.91	0.75
	8,770*	36	19	41	42	194	182	1.1	0.86	1.1
	8,850	66	89	141	75	264	341	2.1	0.88	0.77
	9.830	81	21	126	67	122	277	1.5	1.2	0.44
	9,170	69	38	134	55	276	469	1.9	1.2	0.59
37.	9,835	16	9.8	9.5	24	58	1,350	0.59	0.67	0.04
	9,915	15	11	23	23	71	727	1.5	0.65	0.10
	9,995	35	11	63	32	189	607	1.8	1.1	0.31
	10,075	43	24	61	37	138	573	1.4	1.2	0.24
	10,155	34	12	49	33	161	266	1.4	1.0	0.60
35.	10,700	34	6.9	92	35	184	250	2.7	0.97	0.74
	10,780*	28	25	68	34	175	391	2.4	0.82	0.45
	10,940	34	5.4	50	23	71	372	1.5	1.5	0.19
	11,020	63	27	121	49	192	430	1.9	1.3	0.45
	11,100	47	28	222	58	237	429	4.7	0.81	0.48
	11,180	24	6.0	25	11	53	221	1.0	2.2	0.24
51.	9,490	64	2.0 E	66	56	322	1,000	1.0	1.1	0.32
	9,570	55	1.9 E	130	28	368	1,000	2.4	2.0	0.37
	9,650*	49	1.7 E	37	27	385	1,000	0.76	1.8	0.38
	9,810	34	1.0 E	331	32	389	745	9.7	1.1	0.52
67.	7,310	60	9.5	89	52	292	395	1.5	1.1	0.74
	7,390	80	31	150	68	225	200	1.9	1.2	1.1
	7,470	224	126	258	81	267	388	1.1	2.8	0.69
	7,550	94	16	100	65	166	323	1.1	1.4	0.51
	7,630	81	29	102	55	172	302	1.3	1.5	0.57
	7,710	96	23	157	64	139	243	1.6	1.5	0.57
	7,790*	138	29	62	92	194	332	0.45	1.5	0.58
18.	8,290	71	21	390	110	848	—	5.5	0.65	—
	8,370	73	16	212	68	390	628	2.9	1.1	0.62
	8,450	69	7.9	154	74	317	394	2.2	0.93	0.80
	8,530	49	9.5	135	49	214	443	2.8	1.0	0.48
	8,610	78	13	208	57	442	883	2.7	1.4	0.50
	8,690	68	16	265	85	495	777	3.9	0.80	0.64
	8,770	70	30	287	78	523	827	4.1	0.90	0.63
	8,850	73	21	300	111	694	1,156	4.1	0.66	0.60
	8,930	107	20	499	107	903	1,000	4.7	1.0	0.90

Appendix II
Non-Detrital Trace Element Analysis

Borehole (Depth in Feet)	Ni	Co	Cr	V	Cu	Pb	Ga	Ni/Co	Cr/V	Pb/Ga	Cu/Co
I. Reef Samples											
24. 9,090	26	0.27	3.0	1.8	24	12	10	96	1.7	1.2	89
9,170	50	0.71	9.4	5.0	20	30	19	70	1.9	1.6	28
9,250*	20	0.82	3.2	1.3	11	5.3	14	24	2.5	0.38	13
9,330*	15	0.18	1.5	0.71	—	68	—	83	2.1	—	—
9,410	17	0.14	2.5	0.67	19	16	8.1	121	3.7	2.0	136
9,410	11	0.16	2.2	0.64	16	42	29	69	3.4	1.4	100
33. 9,320*	11	0.56	3.9	2.2	8.8	3.2	2.7	20	1.8	1.2	16
9,480*	9.2	0.23	1.8	1.2	8.1	1.4	0.60 E	40	1.5	2.3	35
9,560*	14	0.68	3.9	5.1	—	—	—	21	0.77	—	—
9,720	9.4	0.24	1.8	2.5	19	1.9	6.4	39	0.72	0.30	79
10,050	15	0.55	4.5	3.1	34	3.9	3.2	27	1.5	1.2	62
10,130*	18	0.54	4.0	3.0	29	25	54	33	1.3	0.46	54
10,290*	17	0.66	3.8	4.0	46	16	41	26	0.95	0.30	70
44. 8,700	24	0.09 E	1.1	3.3	—	—	—	267	0.33	—	—
8,780	16	0.74	2.1	3.2	—	—	—	22	0.66	—	—
8,840	13	1.9	2.3	3.8	—	—	—	6.8	0.60	—	—
9,000*	7.1	0.04 E	0.43	1.3	—	—	—	177	0.33	—	—
9,080*	14	N.D.	1.1	5.6	—	—	—	—	0.20	—	—
9,160	11	0.67	1.6	2.2	—	—	—	16	0.73	—	—
46. 8,700	3.8	0.74	0.90	2.1	—	—	—	5.1	0.43	—	—
8,780*	14	0.21 E	0.21	0.70	—	—	—	78	0.30	—	—
8,860*	4.7	0.05 E	0.59	1.9	—	—	—	94	0.31	—	—
8,940	9.3	1.1	2.1	3.6	22	15	4.3	8.5	0.58	3.5	20
9,020	4.8	0.05 E	0.82	5.5	—	—	—	96	0.15	—	—
9,100	—	—	—	—	24	N.D.	2.0	—	—	—	—
9,180	4.0	0.05 E	0.48	1.5	—	—	—	80	0.32	—	—
9,260*	0.41	0.05 E	0.65	2.3	—	—	—	8.2	0.28	—	—
9,340	26	4.4	0.84	6.8	8.4	17	7.9	5.9	0.12	2.2	1.9
9,400	4.6	1.7	1.3	1.7	—	—	—	2.7	0.76	—	—
32. 9,105	31	1.2	4.1	2.7	46	16	41	26	1.5	0.39	38
9,165*	62	0.96	7.9	1.7	—	—	—	65	4.7	—	—
9,245	7.9	0.65	1.5	4.0	3.2	7.1	9.8	12	0.37	0.72	4.9
9,325	78	2.4	29	18	—	—	—	32	1.6	—	—
9,405	47	0.48	5.4	1.5	—	—	—	98	3.6	—	—
9,485	70	0.49	1.5	0.99	—	—	—	143	1.5	—	—
9,565	90	0.79	5.5	5.5	—	—	—	114	1.0	—	—
9,725	51	0.69	7.7	1.4	4.0	0.88	1.3	74	5.5	0.68	86
9,885	100	2.1	6.5	5.0	2.6	0.58	0.50	48	1.3	1.2	1.2
9,965	13	1.4	1.0	1.4 E	6.0	N.D.	N.D.	9.0	0.71	—	4.3
10,045	8.7	0.65	2.2	1.4	50	3.6	6.5	13	0.16	0.55	77
36. 8,510*	49	0.09 E	5.0	0.3 E	—	—	—	544	17	—	—
8,580	13	0.05 E	1.9	2.0	—	—	—	260	0.95	—	—
8,660	4.9	0.03 E	1.1	1.3	—	—	—	163	0.85	—	—
II. Non-Reef Samples											
11. 9,810*	69	4.0	4.0	9.6	48	71	32	17	0.42	2.2	12
9,890*	96	5.1	5.0	7.7	37	37	35	19	0.65	1.1	7.2
9,970	69	2.5	8.3	6.3	9.0	13	4.7	28	1.3	2.8	3.6
10,130	89	6.8	7.4	21	10	14	7.1	13	0.35	2.0	1.5
10,290	88	2.3	6.0	11	—	—	—	38	0.55	—	—
6. 10,210*	79	3.9	4.2	8.6	—	—	—	20	0.49	—	—
10,290	—	—	—	—	15	21	34	—	—	0.62	—
10,370	37	1.8	5.6	8.8	8.6	20	14	21	0.64	1.4	4.8
10,450	33	4.4	5.6	3.7	11	5.0	5.4	7.5	1.5	0.9	2.5
10,530*	16	4.8	2.7	3.7	13	17	10	3.3	0.73	1.7	2.7
10,610	40	5.4	2.2	5.4	8.4	6.5	7.1	7.4	0.40	0.9	1.6
10,690	67	7.5	6.8	19	19	69	33	8.9	0.36	2.1	2.5
10,770	56	4.4	6.0	13	40	68	9.0	13	0.46	7.6	9.1
10,850	28	2.8	6.2	3.9	26	31	16	10	1.6	1.9	9.3
15. 9,270	16	3.4	3.3	6.1	—	—	—	4.7	0.54	—	—
9,330	39	6.5	11	10	—	—	—	4.6	1.1	—	—
9,410	15	3.5	2.9	5.4	—	—	—	4.3	0.54	—	—
9,490	60	7.5	10	2.7	—	—	—	8.0	3.7	—	—
9,570	19	2.5	1.8	11	—	—	—	7.6	0.16	—	—
13. 8,610*	16	2.2	4.8	5.7	4.1	14	6.6	7.3	0.84	2.1	1.9
8,690	27	5.2	5.3	7.9	47	22	10	5.2	0.67	2.2	9.0
8,770*	34	5.4	5.6	6.9	8.3	15	12	6.3	0.81	1.3	1.5
8,850	12	3.1	2.5	3.2	4.8	15	0.77	3.9	0.78	19	1.5
8,930	19	2.5	3.1	3.6	10	17	4.3	7.6	0.86	4.0	4.0
9,170	7.5	3.7	1.5	7.5	15	20	8.2	2.0	0.20	2.4	4.0

APPENDIX II—(*continued*)

Borehole (Depth in Feet)	*Ni*	*Co*	*Cr*	*V*	*Cu*	*Pb*	*Ga*	*Ni/Co*	*Cr/V*	*Pb/Ga*	*Cu/Co*
12. 8,460	18	4.2	3.8	5.4	20	3.0	5.7	4.3	0.70	0.53	4.8
8,540	14	3.2	5.6	5.7	17	4.2	4.8	4.4	0.98	0.57	5.3
9,620	9.3	2.5	3.2	3.2	7.4	29	3.9	3.7	1.0	7.4	3.0
8,700	9.8	2.6	4.1	4.1	—	—	—	3.8	1.0	—	—
8,780	16	2.6	3.2	3.7	23	36	42	6.2	0.56	0.86	8.8
8,860	7.7	2.3	2.6	3.4	18	50	27	3.3	0.76	1.9	7.8
8,940	35	5.7	6.9	8.5	11	7.9	8.1	6.1	0.81	0.97	1.9
9,020	19	4.1	3.0	4.9	4.6	3.1	2.0	4.6	0.61	1.5	1.1
9,100	18	3.0	3.9	3.9	12	17	13	6.0	1.0	1.3	4.0
5. 9,080	8.1	1.5	1.9	2.3	—	—	—	5.4	0.83	—	—
9,160	22	5.4	2.3	7.0	—	—	—	4.1	0.33	—	—
9,240	20	4.6	3.5	4.5	—	—	—	4.3	0.78	—	—
9,320	27	4.4	4.1	3.9	—	—	—	6.1	1.1	—	—
9,400	6.2	1.6	1.3	1.3	—	—	—	3.9	1.0	—	—
37. 9,835*	7.1	0.64	3.0	4.5	—	—	—	11	0.67	—	—
9,915*	21	1.5	4.3	4.7	—	—	—	14	0.91	—	—
9,995*	18	1.6	3.2	2.9	—	—	—	11	1.1	—	—
10,075*	21	2.5	3.1	2.6	—	—	—	8.4	1.2	—	—
10,155*	8.5	2.0	1.5	1.8	—	—	—	4.2	0.83	—	—
35. 10,700*	6.2	0.18	1.2	1.5	6.6	5.0	3.1	34	0.80	1.6	37
10,780	15	2.3	2.3	8.0	5.5	1.1	0.52	6.5	0.29	2.1	2.4
10,940	9.8	1.9	1.3	2.5	—	—	—	5.2	0.52	—	—
11,020	19	3.5	3.0	9.0	16	15	13	5.4	0.33	1.2	4.6
11,100	28	2.9	4.6	10	27	13	9.5	9.7	0.46	1.4	9.3
11,180*	22	4.6	2.8	14	19	17	29	4.8	0.20	0.59	4.1
11,260	23	5.1	3.7	12	20	5.2	4.0	4.5	0.31	1.3	3.9
58. 9,810	—	—	—	—	23	29	17	—	—	1.7	—
9,890	—	—	—	—	27	52	29	—	—	1.8	—
9,970	—	—	—	—	21	7.8	1.4 E	—	—	5.6	—
10,050	—	—	—	—	40	44	0.5 E	—	—	88	—
10,130	—	—	—	—	7	5.1	15	—	—	0.34	—
10,210	—	—	—	—	20	2.9	8.9	—	—	0.33	—
51. 9,490	38	3.2	6.0	6.5	15	6.1	4.3	12	0.92	1.4	4.7
9,570	57	2.2	9.6	9.3	10	5.3	N.D.	26	1.0	—	4.5
9,650*	45	4.3	7.9	13	—	—	—	10	0.61	—	—
9,810	—	—	—	—	6.1	9.5	5.7	—	—	1.7	—
9,890	—	—	—	—	13	41	10	—	—	4.1	—
9,970	—	—	—	—	30	25	46	—	—	0.54	—
10,050	—	—	—	—	17	37	6.3	—	—	5.9	—
67. 7,310	18	4.0	4.3	5.8	18	25	32	4.5	0.74	0.78	4.5
7,390	48	9.5	7.6	13	21	8.3	11	5.0	0.58	0.75	2.2
7,470*	35	9.2	6.2	15	19	19	5.1	3.8	0.41	3.7	2.1
7,550	34	5.1	6.6	4.5	21	7.9	5.2	6.7	1.5	1.5	4.1
7,630*	4.9	7.4	7.8	8.7	36	19	18	0.66	0.90	1.7	4.9
7,710	29	11	6.4	15	15	20	18	2.6	0.43	1.1	1.4
7,790	23	6.8	6.3	11	10	32	28	3.4	0.57	1.1	1.5
61. 7,000	30	18	7.3	0.6 E	8.1	12	7.1	17	12	1.7	0.45
7,080	31	2.0	2.1	4.2	15	15	N.D.	15	0.50	—	7.5
7,160*	22	7.4	3.6	8.2	34	25	13	3.0	0.44	1.9	4.6
7,240	42	6.8	3.9	12	18	19	18	6.2	0.32	1.1	2.6
7,320	15	6.2	6.6	13	41	34	23	2.4	0.51	1.5	6.6
7,400*	67	13	11	36	88	96	55	5.2	0.30	1.7	6.8
7,450	33	9.3	16	12	19	31	9.6	3.5	1.3	3.2	2.0
7,560	21	9.1	7.0	11	14	38	5.3	2.3	0.64	7.2	1.5
7,640*	28	8.6	4.4	11	17	9.2	9.6	3.3	0.39	0.96	2.0
7,720*	37	15	7.9	20	33	58	19	2.5	0.39	3.1	2.2
7,800*	20	5.3	10	8.7	25	18	14	3.8	1.1	1.3	4.7
7,880*	13	0.20 E	0.11 E	7.0	—	—	—	65	0.01	—	—
7,960	48	4.0	3.0	7.0	8.6	26	7.9	12	0.43	3.3	2.1
23. 8,900*	99	8.6	16	11	—	—	—	11	1.5	—	—
8,980	29	6.4	7.6	11	—	—	—	4.5	0.69	—	—
9,060*	29	5.5	8.3	2.0	—	—	—	5.3	4.1	—	—
9,140	39	8.2	5.2	10	—	—	—	4.8	0.52	—	—
9,220	46	6.0	7.6	19	—	—	—	7.7	0.40	—	—
9,300*	32	0.85 E	5.8	3.4 E	—	—	—	38	1.7	—	—
9,380*	140	0.67 E	13	2.7 E	—	—	—	209	4.8	—	—
18. 8,290	31	6.9	10	19	—	—	—	4.5	0.53	—	—
8,370	38	9.3	13	15	21	4.0	10	4.1	0.87	0.40	2.3
8,450	33	12	11	20	13	28	8.0	2.7	0.55	3.5	1.1
8,530	14	4.6	4.6	6.8	18	29	33	3.1	0.68	0.88	3.9
8,610	31	8.7	5.3	13	45	10	6.7	3.6	0.41	1.5	5.2
8,690	26	6.8	8.2	12	26	9.4	0.35 E	3.8	0.68	27	3.8
8,770	24	6.9	6.8	11	23	7.9	4.0	3.5	0.62	2.0	3.3
8,850	32	11	6.2	10	23	33	15	2.9	0.62	2.2	2.1
8,930	58	11	7.6	13	12	14	13	5.3	0.58	1.1	1.1

Appendix III

Borehole Reference Numbers

Borehole No.	*Reference No.*	*Name*
24	15-8-67-22-W5M	Gulf Little Smoky
11	10-32-63-21-W5M	Pan American Tony Creek A1
6	10-27-62-21-W5M	Phillips Tony A
4	12-9-61-21-W5M	Pan American Big Stone Creek A1
13	7-22-64-19-W5M	Phillips Kaybob A
12	12-6-64-12-W5M	Imperial Burntwood
5	5-35-62-18-W5M	Cal. Std. Gulf Kaybob
33	14-13-68-25-W5M	Amerada Crown XF
37	7-19-69-25-W5M	Amerada Crown AD
35	4-8-68-1-W5M	Amerada Crown AN
44	10-13-70-25-W5M	Amerada Crown VF
58	6-11-73-26-W5M	Amerada Crown GF
51	4-28-72-25-W5M	B.A. Crooked Creek No. 1
46	3-11-71-24-W5M	Amerada Crown UF
67	14-15-75-22-W5M	Imperial Little Smoky No. 1
61	16-6-74-20-W5M	B.A. Little Smoky
32	1-8-68-23-W5M	Amerada Crown AJ
36	1-12-69-22-W5M	Shell South Sturgeon Lake SB
23	15-28-67-21-W5M	Gulf Tosegun
18	6-35-66-19-W5M	Gulf Goose River

References

Graf, D. L., 1960, Geochemistry of carbonate sediments and sedimentary carbonate rocks: Ill. State Geol. Survey Circ. 301.

Hirst, D. M., 1962, The geochemistry of modern sediments from the Gulf of Paria—II, The location and distribution of trace elements: Geochim. et Cosmochim. Acta, v. 26, p. 1147–1187.

——— and Nicholls, G. D., 1958, Techniques in sedimentary geochemistry. (1) Separation of the detrital and non-detrital fractions of limestones: Jour. Sed. Petrology, v. 28, 461–468.

Humphries, R. G., and Hage, C. O., June, 1955, Geology, development and reserves, Sturgeon Lake oil fields, Alberta: Canadian Mining and Metall. Bull.

Ingerson, E., 1962, Problems of the geochemistry of sedimentary carbonate rocks: Geochim. et Cosmochim. Acta, v. 26, p. 815–847.

Keith, M. L., and Degens, E. T., 1959, Geochemical indicators of marine and fresh-water sediments, *in* Researches in geochemistry: p. 38–61, John Wiley, N. Y.

Krynine, P. D., 1948, The megascopic study and field classification of sedimentary rocks: Jour. Geology, v. 56, p. 130–165.

Nicholls, G. D., 1958, Sedimentary geochemistry: Petroleum, Sept., 1958.

Nicholls, G. D., and Loring, D. H., 1962, The geochemistry of some British Carboniferous sediments: Geochim. et Cosmochim. Acta, v. 26, p. 181–223.

Potter, E. D., Shimp, B. F., and Witters, J., 1963, Trace elements in marine and fresh-water argillaceous sediments: Geochim. et Cosmochim. Acta, v. 27, p. 669–694.

Shaw, D. M., and Bankier, J. D., 1954, Statistical methods applied to geochemistry: Geochim. et Cosmochim. Acta, v. 5, p. 111–123.

Weber, J. N., and Smith, F. G., 1961, Rapid determination of calcite-dolomite ratios in sedimentary rocks: Jour. Sed. Petrology, v. 31, p. 130–131.

Reprinted from
BULLETIN OF THE AMERICAN ASSOCIATION OF PETROLEUM GEOLOGISTS
VOL. 50, NO. 10 (OCTOBER, 1966), PP. 2068-2081, 14 FIGS.

STRATIGRAPHIC *VS.* STRUCTURAL CONTROLS ON CARBONATE-MOUND HYDROCARBON ACCUMULATION, ANETH AREA, PARADOX BASIN[1]

JAMES A. PETERSON[2]
Missoula, Montana

ABSTRACT

Pennsylvanian oil and gas accumulations in the southern Paradox basin are in carbonate mounds of Desmoinesian age. Major oil production at Aneth, Ismay, Tohonadla, Gothic Mesa, Anido Creek, and other fields is from algal mounds elongate in a general northwest-southeast direction along the basin shelf. Carbonate reservoirs are associated closely with sapropelic black shale and evaporite, which occur in cyclic repetition in the shelf area and grade basinward to a predominantly salt section.

All reservoirs appear to be isolated bodies of porous carbonate, mostly limestone. About 30 oil and gas fields productive from the Pennsylvanian have been found in the Four Corners area; approximately half are classified as stratigraphic and the other half as either structural or structural-stratigraphic. In almost all fields, it can be demonstrated that the accumulation would have formed even if no structural closure were present, although in many fields the oil is localized by subsequent structural growth.

The Aneth, Ismay, and Cache fields are primary examples of Paradox basin fields showing major stratigraphic and only minor structural influence on accumulation.

Introduction

Oil production in the Aneth area of southeastern Utah and southwestern Colorado is from isolated carbonate-mound reservoirs that provide the major trapping mechanism for accumulation. Structural influence is local, within mound trends, some of which are only 1–2 mi. wide, but may extend for several miles in an elongate, generally northwest-southeast direction.

Historically, the Four Corners area has undergone sporadic exploration activity since the early 1900s. The initial play was in the Paradox basin where the first oil was found in 1907 at the shallow Mexican Hat field along the San Juan River on the Monument upwarp. Pennsylvanian gas subsequently was found at several small- to medium-size fields south and southwest of the main Aneth complex (Fig. 1). Most of the major fields were discovered in 1956 and 1957, and several small fields were discovered during the intensive drilling program following discovery of Aneth in 1956. Total estimated ultimate oil reserves of the area are 300–400 million bbls.

Two main carbonate-mound intervals, or "cycles," known in the industry as the Desert Creek and Ismay "zones," are of economic interest (Fig. 2). Porosity development within these two "zones" occurs basinward from the main Pennsylvanian shelf-carbonate belt along the southwest flank of the Paradox basin. The relatively basinward position and the associated lateral facies relations are of great importance in providing the prominent isolated stratigraphic trap in the carbonate-mound oil accumulations of the Aneth area.

The purpose of the present paper is to analyze the local stratigraphic and structural conditions influencing entrapment of oil in the Aneth area. Several other papers, listed in the references, provide a more detailed discussion of the regional stratigraphy and structure and of the lithofacies-porosity aspects of shelf-carbonate mounds in the Paradox basin.

General Geologic Setting

The Paradox basin is within the central part of the Colorado Plateau physiographic province. Regionally, the basin is bounded on the south by the Four Corners platform, the Defiance uplift, and the Monument upwarp; on the west by the Circle Cliffs uplift and the San Rafael swell; and on the east by the Uncompahgre uplift. All these major

[1] Read before the Rocky Mountain Section of the Association at Billings, Montana, September 28, 1965. Manuscript received, January 28, 1966; accepted, April 22, 1966.

[2] Department of Geology, University of Montana. Acknowledgments are made with appreciation to the Shell Oil Company for use of information and files in developing the ideas presented here and to Shell geologists and other Paradox basin stratigraphers for fruitful discussions of stratigraphy and sedimentation in the Aneth area. Special thanks are due D. P. McGookey and C. W. Spencer, Texaco Inc., and W. L. Stokes, University of Utah, for many worthwhile suggestions during critical review of the manuscript.

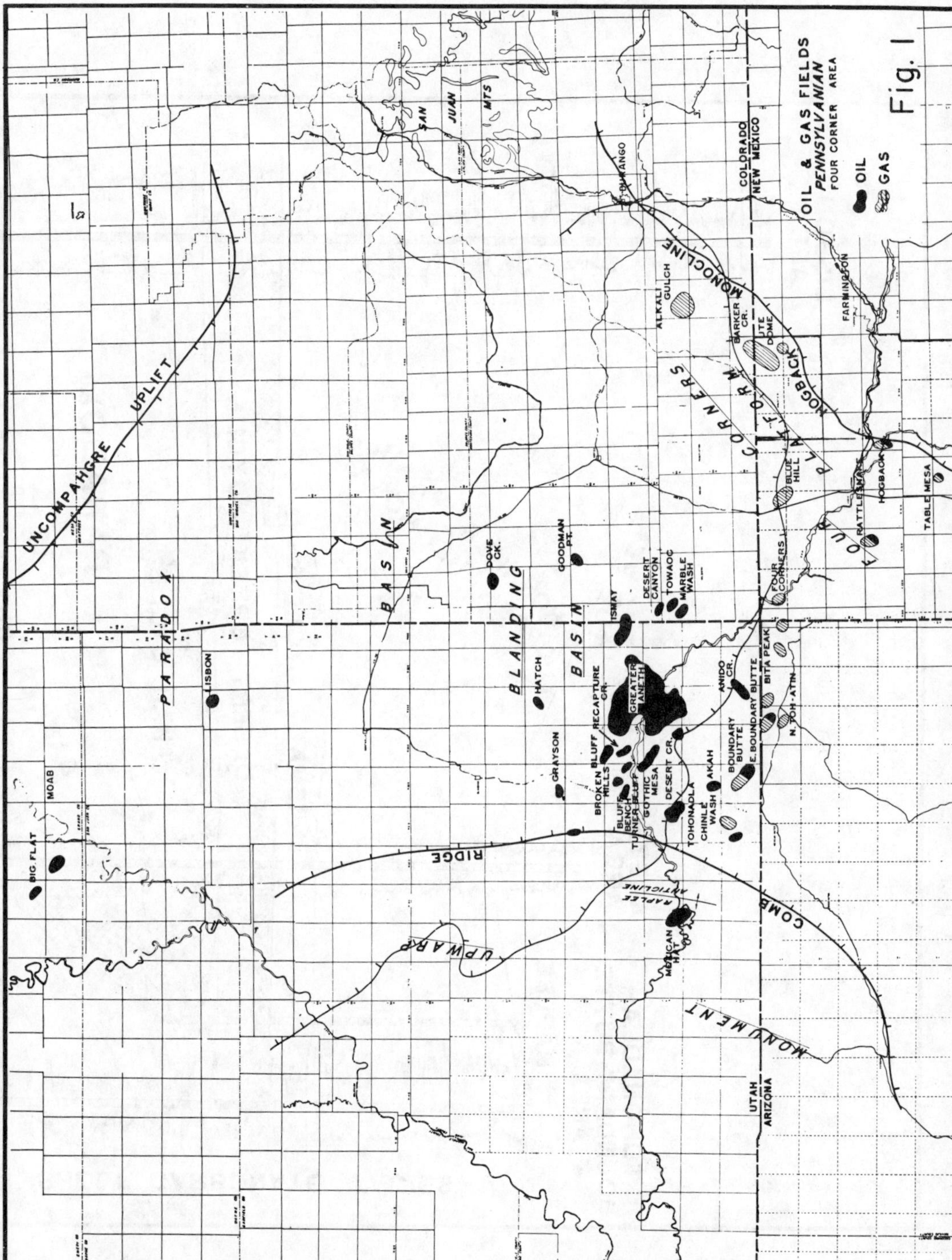

FIG. 1.—Index map showing Pennsylvanian oil and gas fields and major structural features of Paradox basin. Four Corners area (Utah, Colorado, New Mexico, and Arizona).

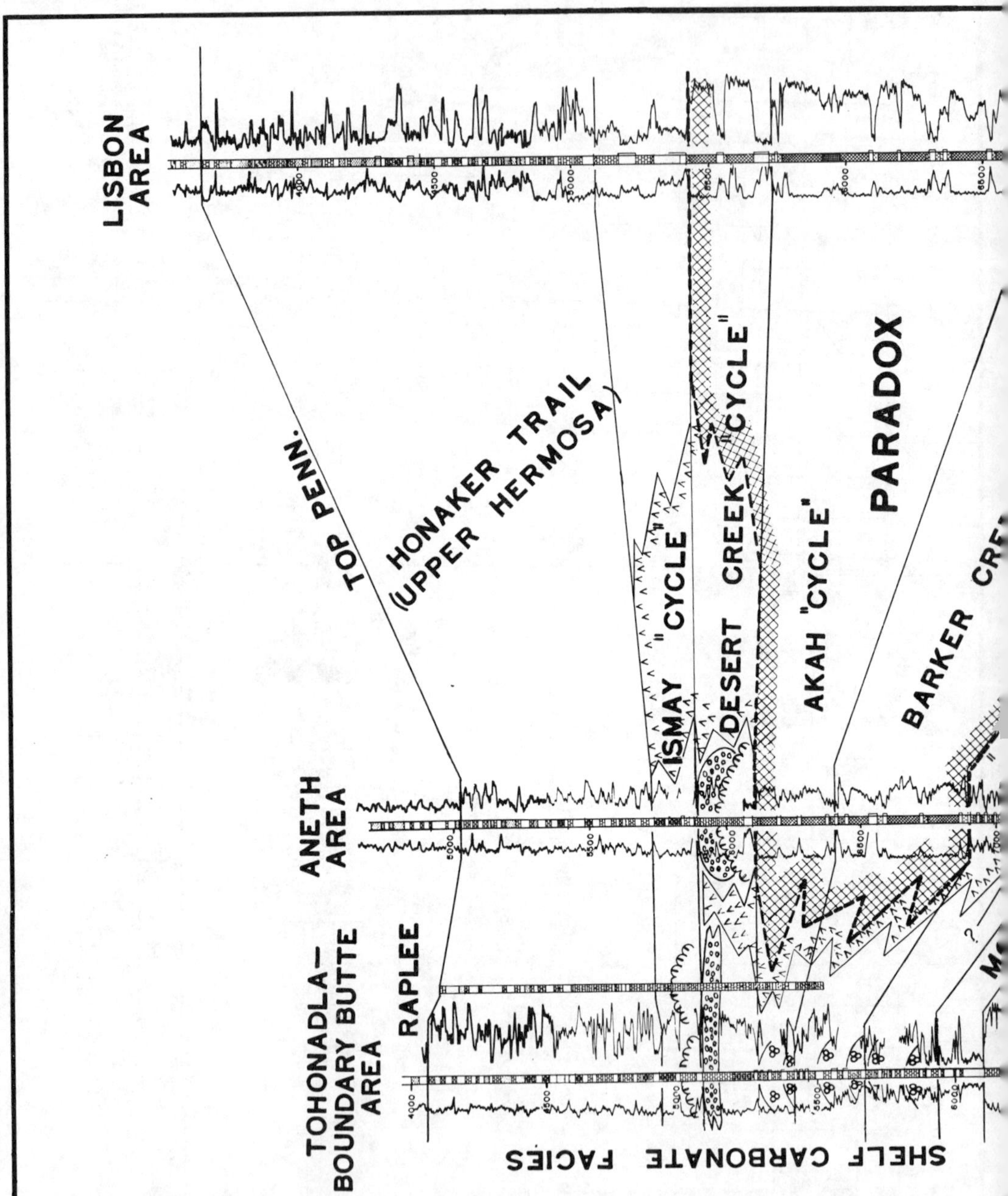
LISBON AREA
ANETH AREA
TOHONADLA—BOUNDARY BUTTE AREA
RAPLEE
TOP PENN.
HONAKER TRAIL (UPPER HERMOSA)
ISMAY "CYCLE"
DESERT CREEK "CYCLE"
AKAH "CYCLE"
PARADOX
BARKER
SHELF CARBONATE FACIES

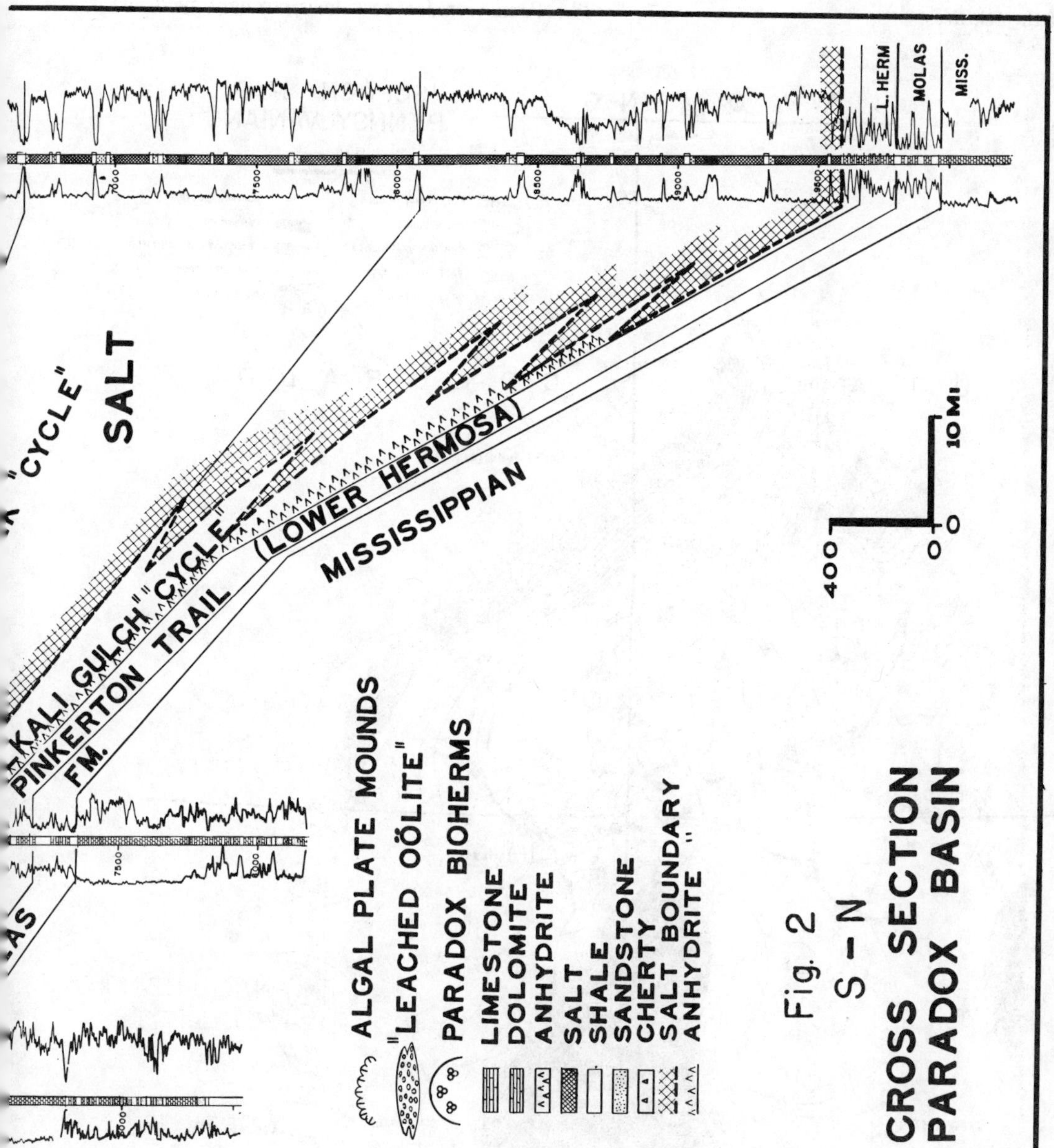

FIG. 2.—South-north cross section of Pennsylvanian in Paradox basin, showing relations between shelf carbonate-mound reservoir facies and evaporite facies.

features underwent several early stages of tectonic growth dating as long ago as Pennsylvanian time, with probable lesser earlier development of some of the features.

Regional distribution of the important Pennsylvanian stratigraphic units is shown in Figure 2. The thickness distribution is shown in Figure 3. The Pennsylvanian section is strongly cyclic with

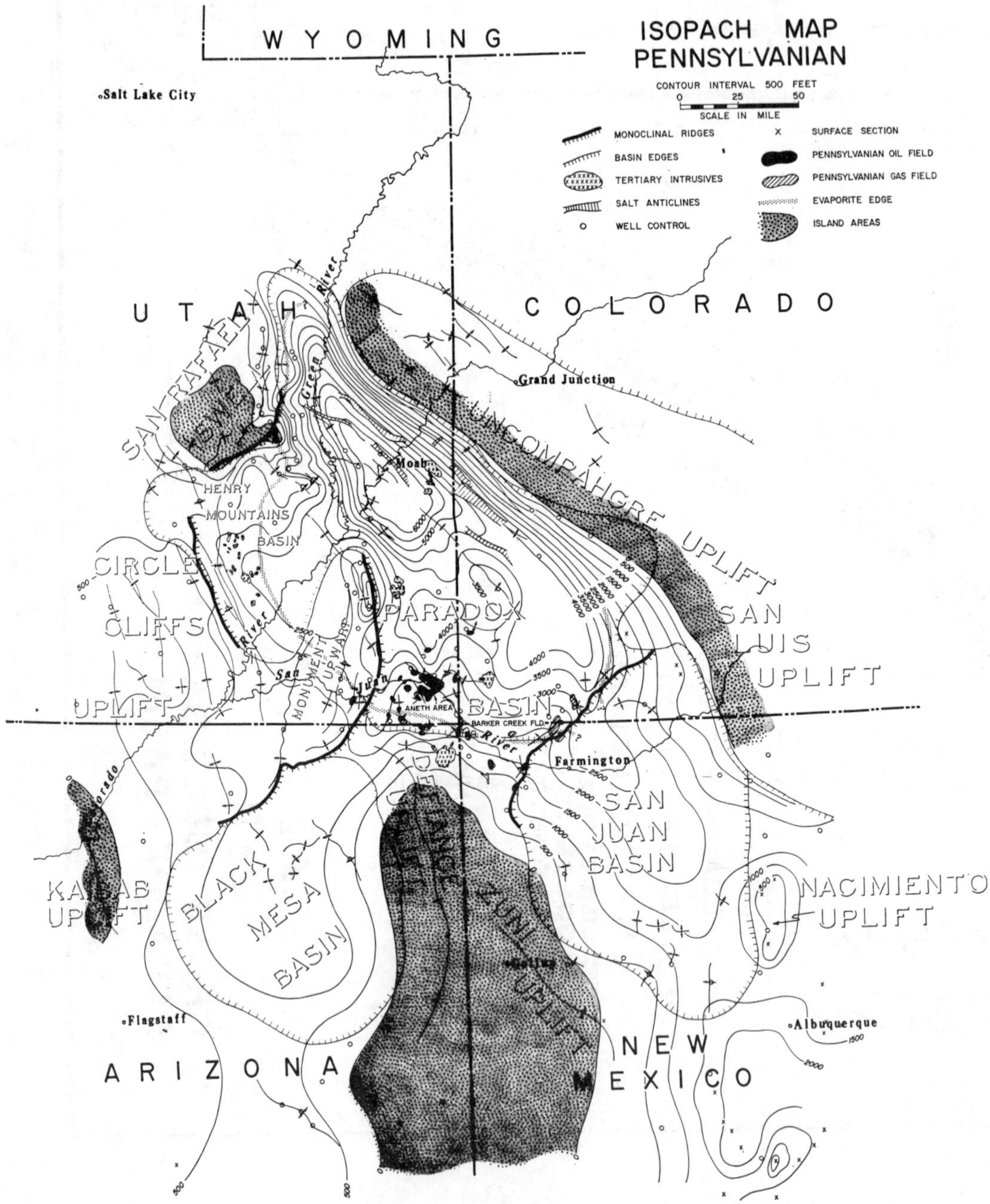

Fig. 3.—Isopachous map, Pennsylvanian System, Four Corners area (after Peterson, 1959). Edge of Paradox basin evaporites shown by stippled band. Contour interval = 500 ft.

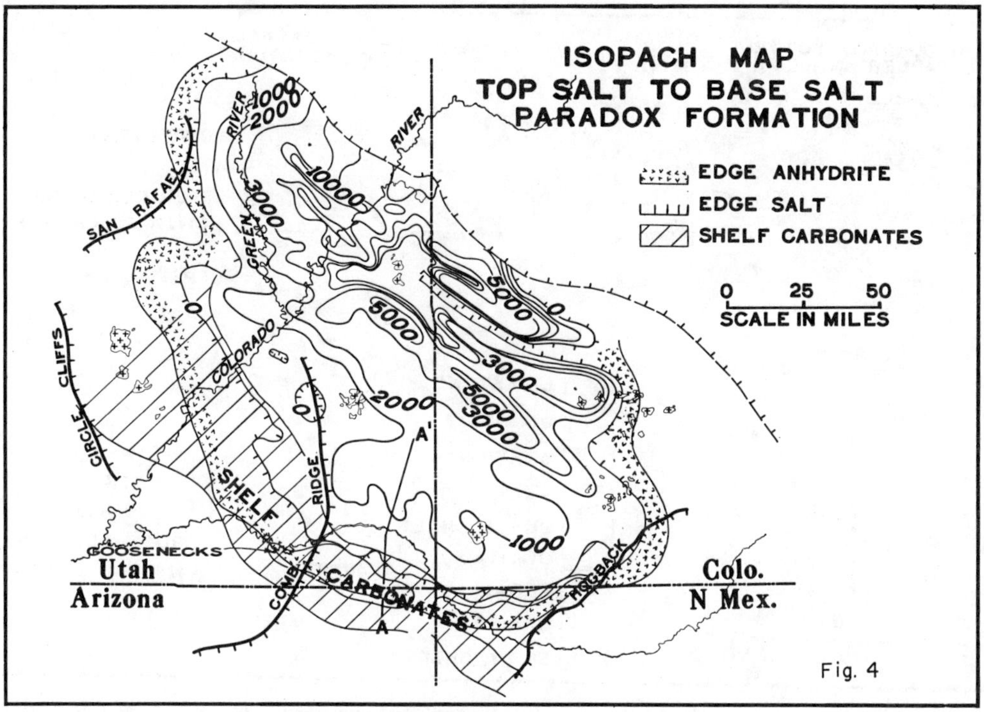

FIG. 4.—Isopachous map, Paradox salt, showing relation with age-equivalent shelf carbonates. Contour interval = 1,000 ft. Line of cross section of Figure 5 indicated by A-A′.

each cycle bounded by basin-wide black sapropelic "shale" marker beds composed mostly of carbonate and black organic material with a relatively low clay or silt content. These beds are defined more correctly as carbonaceous dolomite but commonly are referred to as "shale" because of their prominent radioactive character on gamma ray-neutron logs. The main part of the basin contains thick beds of halite with thin units of anhydrite, black shale, and dolomite, repeated cyclically within the main body of the Paradox Formation. This unit is more than 5,000 ft. thick within the central basin area where the salt is thickened locally to more than 10,000 ft. by flowage in the northwest-southeast-trending diapiric structures of the salt-anticline belt (Fig. 4).

Individual cycles grade from evaporites in the central basin to a shelf-carbonate-mound facies along the southwest flank of the basin. Most of the individual shelf-carbonate-mound trends of the more than 25 Paradox evaporite cycles developed repeatedly in a generally coincident belt along the basin shelf. However, the mound belts of the two youngest major Paradox cycles (the Desert Creek and Ismay cycles) are developed basinward from the main shelf-carbonate belt (Fig. 2). Each of these two cycles is composed of two or more subcycles and together contain the broadest and thickest productive mound belts of the Aneth area. The main rock types and general facies relations of the Desert Creek and Ismay cycles are shown in the cross section (Fig. 5).

Desert Creek Cycle

Porosity in the Desert Creek cycle is in two reservoir types, a calcareous "algal leaf" limestone-mound development in the *lower* part, and a "leached oölite" limestone and dolomite interval in the *upper* part (Fig. 5). The algal reservoir at Aneth is slightly over 100 ft. thick in the central part of the mound, and the leached oölite averages approximately 100 ft. in most of the bank area, giving a maximum mound thickness of just over 200 ft. The overall porous mound buildup is slightly "horseshoe-shaped" with the major breadth and thickness in the greater Aneth field area and two relatively narrow and thin irregular arms of algal mound extending northwest

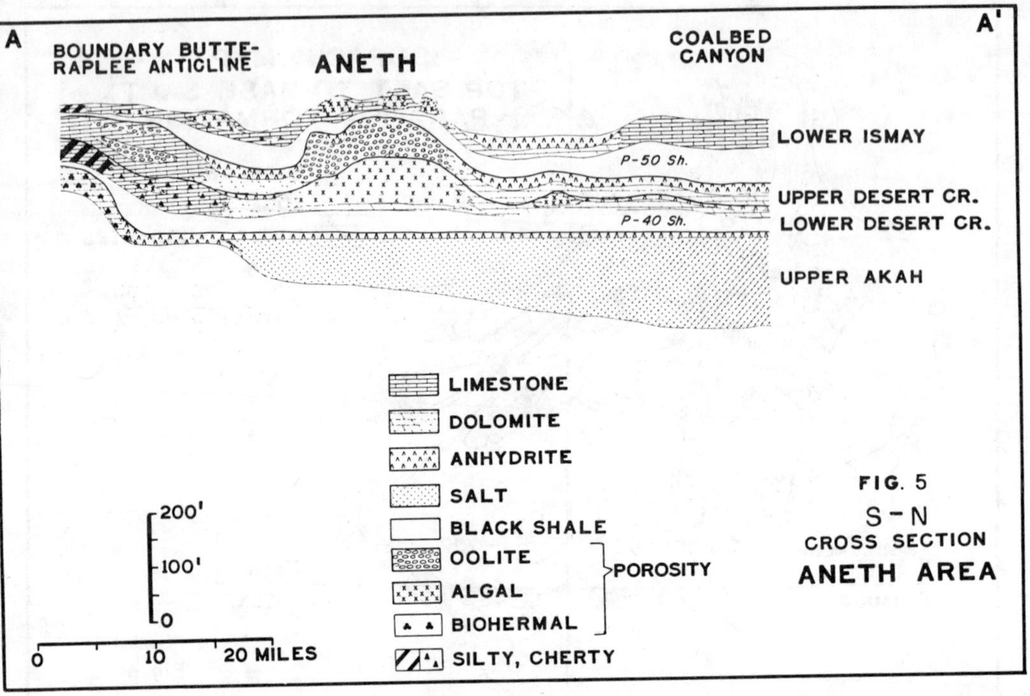

FIG. 5.—South-north cross section A-A′ across Aneth bank showing details of rock-facies interrelations. Location of section A-A′ is shown in Figure 4.

through the Bluff and Gothic Mesa fields. Figure 6 illustrates the general shape and extent of the Desert Creek mounds and also shows the highly isolated position of the Aneth area porosity, enclosed by the evaporite facies. Figures 7 and 8 are photomicrographs of the *lower* Desert Creek algal reservoir rock from the southwest part of the Aneth bank where the original nature of the sediment is relatively well preserved. The porous rock is formed largely of calcified remains of the alga *Ivanovia* (Parks, 1958; Wray and Konishi, 1960; Murray, 1960) intermixed with pelleted mud lenses which probably represent pockets of fine limestone mud trapped during mound growth. Fusulinids, small Foraminifera, some brachiopods, and *Chaetetes* corals are rare in the algal-mound rock. Porosity is almost universally associated directly with pockets of maximum accumulation of leafy *Ivanovia* material although the fine details of algal structures generally are destroyed by recrystallization.

The *upper* Desert Creek leached oölite facies consists largely of oölites, pellets, and coated fossil fragments that have been irregularly leached and recrystallized, and partly dolomitized (Figs. 9, 10). Oölitic and pelletal rocks generally are thicker on the borders of the mound. The inner part of the mound contains beds of bioclastic and recrystallized limestone mud and the characteristic porous oölite-pelletal beds which are discontinuous in an apparent channel-like pattern. Porosity is associated generally with leaching and partial dolomitization of the oölite and pellet-rock facies and is on the average *higher* than in the underlying algal facies. Permeability of the leached oölite reservoir rock averages considerably *lower* than that of the algal reservoir. The upper oölite beds attain maximum thickness in the main mound area of the Aneth field, where the beds are more than half of the porous section. The oölite interval changes markedly to chalky dolomite and anhydrite facies away from the Aneth field; the oölite interval is not in the other Desert Creek cycle fields of the area (Fig. 6). A leached oölite facies also is in the *upper* Desert Creek in a broad belt along the southwest basin shelf (Fig. 6) where it commonly contains as much as 50 ft. or more of highly porous oil-saturated limestone. However, along this extensive belt the Desert Creek section has not yet been found with sufficient permeability for substantial economic production.

Ismay Cycle

Distribution of limestone-mound porosity in the Ismay cycle is illustrated by Figure 11. In most of the area, porosity in this unit is largely in the *lower* Ismay subcycle. Geometry of the porous bodies is characterized by generally northwest-southeast, somewhat discontinuous, elongate, irregularly shaped, narrow belts of algal limestone banks. The reservoir rock is similar to that of the algal and pelleted-mud reservoir rock of the *lower* Desert Creek cycle (Fig. 7). A field example of a *lower* Ismay reservoir is in the photograph (Fig. 12) taken in the San Juan River canyon east of Mexican Hat, Utah. The photograph illustrates local thickness changes within a belt of mound development. Conceivably some of the subsurface anomalies in the area can be explained by patterns similar to those in the San Juan River exposures. A well drilled in the main mound trend

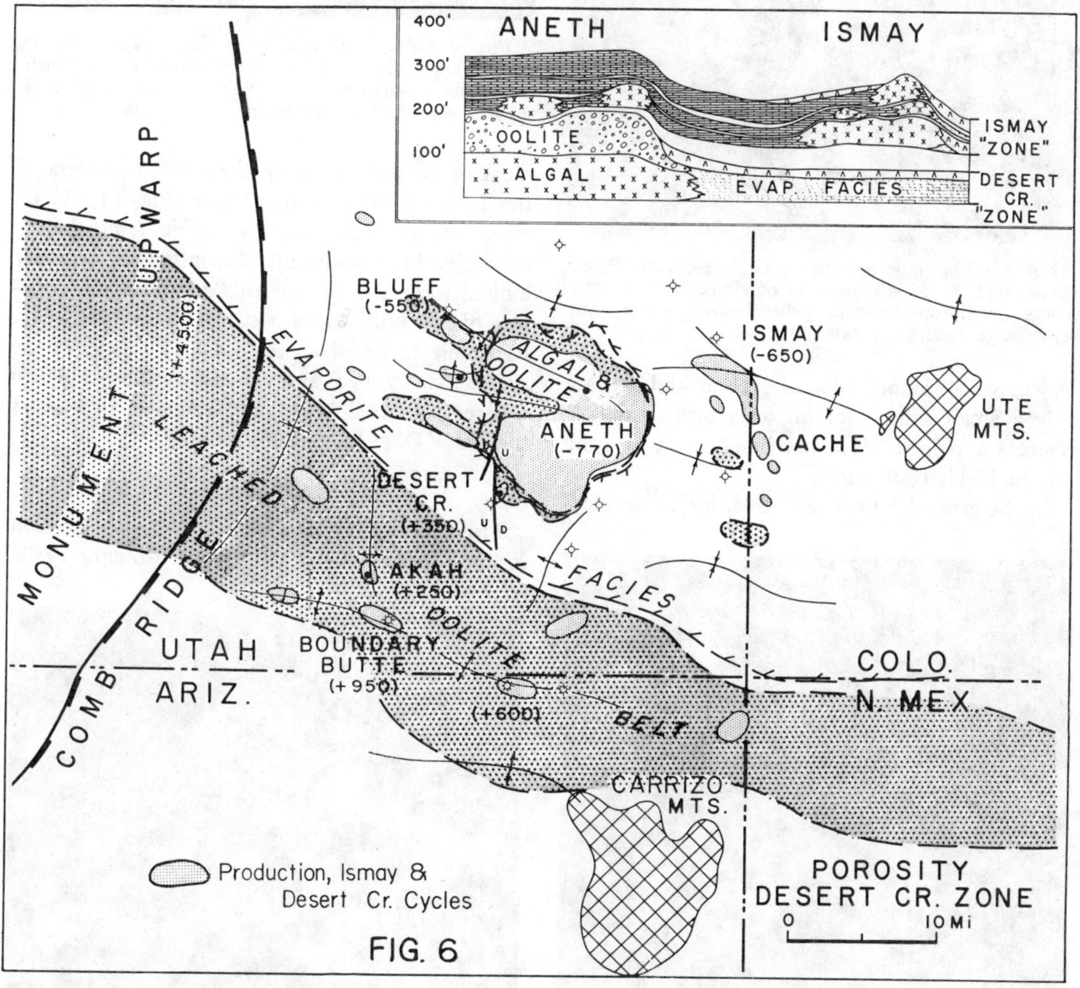

Fig. 6.—Main porous facies of Desert Creek cycle, Aneth area (finely stippled pattern). Leached oölite porosity is only in upper Desert Creek interval and, in Aneth area, is only in Aneth field as shown on schematic west-east cross section at upper right. Evaporite facies of dolomite, anhydrite, and black shale (shown by diagonal hachured line) coincides with limit of upper Desert Creek oölite-bank distribution. Carrizo Mountains and Ute Mountain are Tertiary intrusives.

FIG. 7.—Photomicrograph of algal reservoir, Desert Creek field. Rock is composed of elongate algal fragments of fibrous calcite, pelletal mud lenses, and some large calcite crystals.

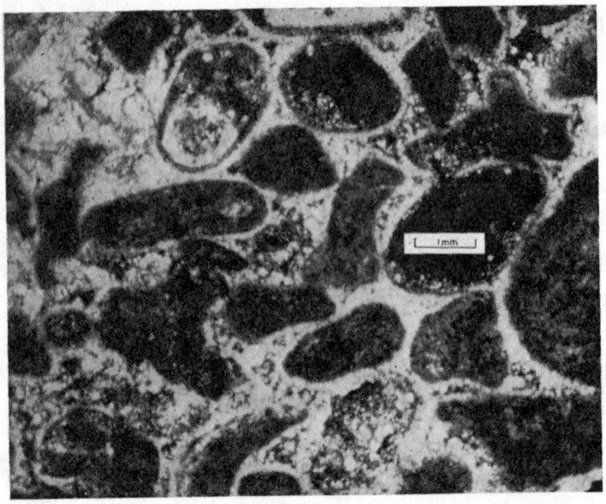

FIG. 9.—Photomicrograph of bank-edge "leached oölite" reservoir of upper Desert Creek cycle, southwest side of greater Aneth field. Matrix material is mostly recrystallized limestone mud.

may have a thinned mound section and recover only a small amount of oil on a drill-stem test, whereas a well drilled a few hundred yards away may be highly commercial.

In the general Ismay field area, major development of porosity is along the eastern borders of the Ismay mound buildup trend (Fig. 14). Here, reservoirs are best developed in the *lower* Ismay subcycle, but significant discontinuous porosity buildups are also in the *middle* and *upper* subcycles of the unit (cross sections, Figs. 6, 11, 13). Reservoir facies of all three Ismay subcycles are dominated by algal and pelleted-mud rocks which are recrystallized in a manner similar to that of the *lower* Desert Creek mounds.

GENESIS OF PARADOX CYCLES

Most of the Pennsylvanian carbonate section in the Paradox basin displays a strongly cyclic

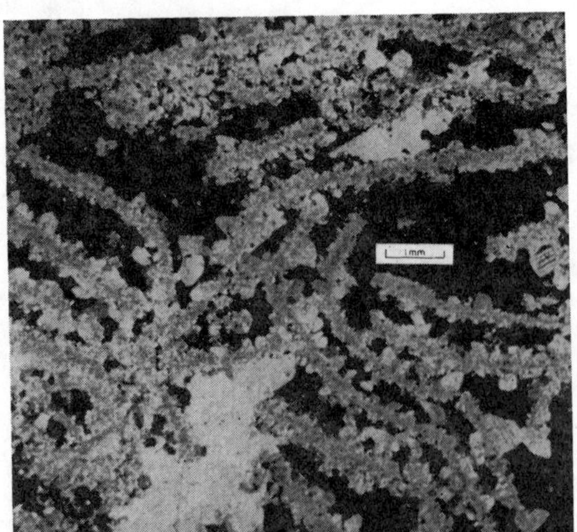

FIG. 8.—Photomicrograph of highly recrystallized algal reservoir, lower Desert Creek cycle, Aneth field. Pore spaces are black.

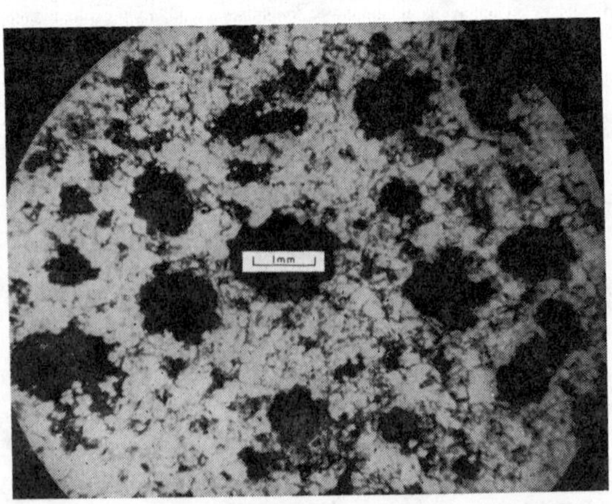

FIG. 10.—Photomicrograph of highly recrystallized and partly dolomitized "leached oölite" reservoir of upper Desert Creek cycle center of greater Aneth field. Pore spaces are black.

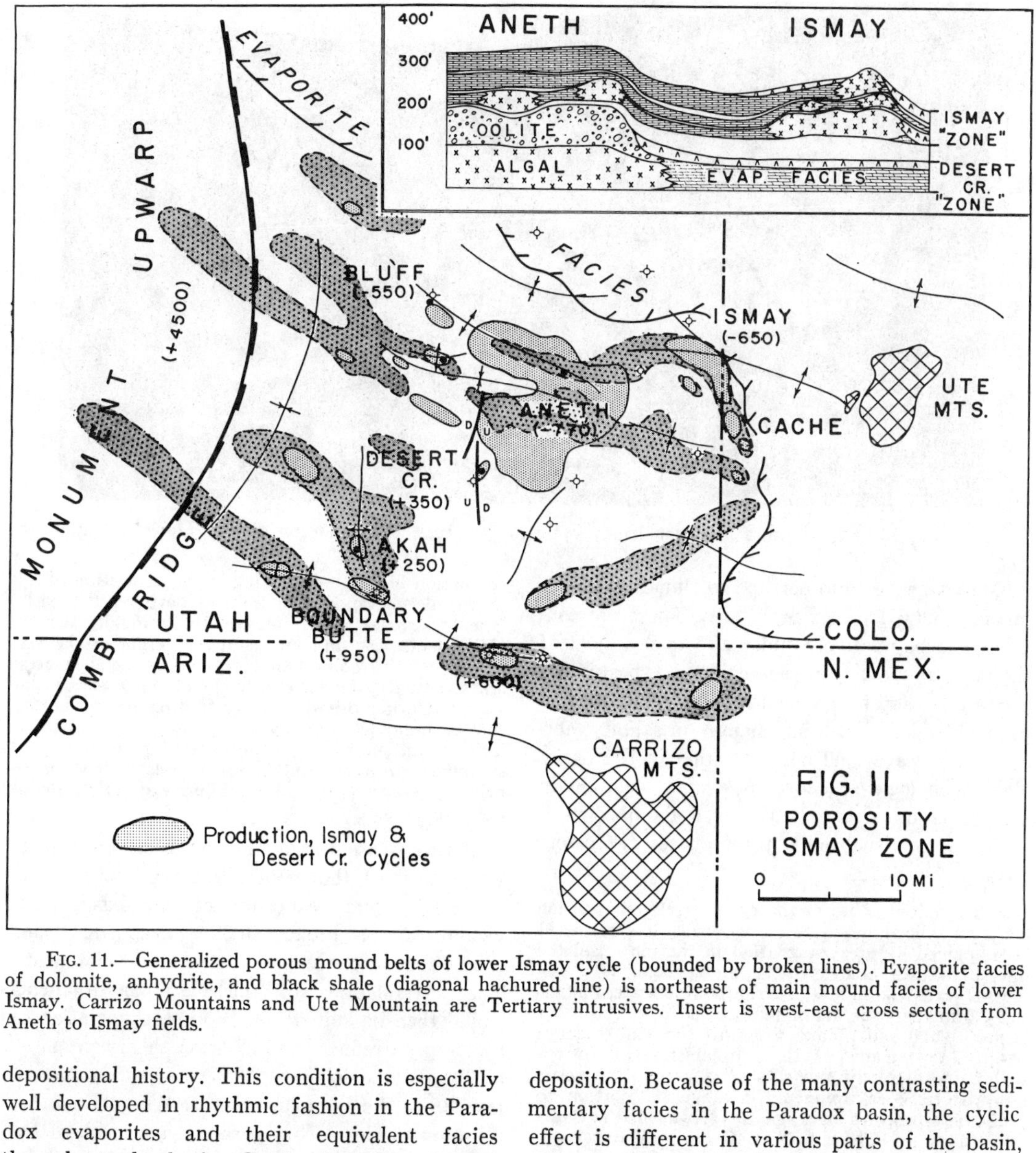

FIG. 11.—Generalized porous mound belts of lower Ismay cycle (bounded by broken lines). Evaporite facies of dolomite, anhydrite, and black shale (diagonal hachured line) is northeast of main mound facies of lower Ismay. Carrizo Mountains and Ute Mountain are Tertiary intrusives. Insert is west-east cross section from Aneth to Ismay fields.

depositional history. This condition is especially well developed in rhythmic fashion in the Paradox evaporites and their equivalent facies throughout the basin. Cyclic deposition is well known in Pennsylvanian sediments throughout the North American and European continents, but their cyclic nature is not in every place easy to recognize unless unusual and variable rock types are within the sequence (*e.g.*, coal, underclay, *etc.*, in the Mid-Continent United States). In the Paradox basin the predominance of salt, anhydrite, and black shale (*i.e.*, well-developed evaporitic conditions) in repetitional sequences demonstrates the strongly rhythmic nature of the deposition. Because of the many contrasting sedimentary facies in the Paradox basin, the cyclic effect is different in various parts of the basin, *e.g.*, on the carbonate shelf, in the central basin evaporitic facies, and in the arkosic facies along the east side of the basin. The orderly repetition of the cyclic layers is thought to be related to sea-level changes probably caused by cycles of Carboniferous glaciation in the southern hemisphere, for which more evidence is accumulating. The development and the exact type of sediment in each cycle also are related to local changes in rate and type of clastic influx and its influence on submarine topography, to changes in rate of tec-

FIG. 12.—Exposed algal mound of lower Ismay cycle in San Juan River canyon east of Mexican Hat, Utah.

tonic movements, and perhaps to climatic changes in the general Paradox basin area. These are considered to be relatively minor effects and most of the vertical changes in the sediment types within each cycle can be explained best by sea-level fluctuations and the resulting changes in salinity, aeration, and wave- and current-action in the basin. The chronologic sequence of events resulting in the formation of a well-developed Paradox basin cycle can be visualized in three phases (Peterson, 1959, p. 506–511).

1. The initial phase of the cycle is related to rising sea level following a time of regionally low sea level accompanied by great stagnation in the more isolated inter-island basinal areas of the 'archipelago.' With rising sea level, the inflow of fresh marine waters intensifies, bringing an influx of marine plankton and other pelagic plant and animal life into the still-stagnant basin. The remains of these organisms, intermixed with some fine clay and silty, clastic material, settle into the toxic environment of the basin bottom to form the sapropelic black 'shales' typical of the Paradox formation. These beds grade to red and green shales and eventually to coarser clastics toward the major uplift areas.

2. Continued freshening of waters occurs with further rise of sea level. Inflowing currents become stronger and more consistent. The waters lap farther up on the island flanks, and in some of the cycles current systems may have been set up that completely traversed the basins, especially across the marginal shelf areas. Freshened and well-aerated marine waters dominate the shallow water shelf margins where biostromal banks and small reefs develop. These grade basinward into dolomites and sapropelic beds and to sandy carbonates toward the adjacent uplifts.

3. Sea level begins to lower. Another phase of basin stagnation begins, culminating in the deposition of salt in the deeper and more isolated parts of the basin. The biostromal-biohermal growths gradually die off. With shallowing and increased restriction of the waters, pelletal or oolite banks develop in the shelf areas and eventually the skeletal banks or bioherms become awash. During this phase, lithification of the porous skeletal buildups probably occurs along with recrystallization and possibly dolomitization, development of secondary porosity, and in some cases destruction of primary porosity by pore infillings of anhydrite or dolomite crystals.

During phase three (the lower sea-level stage when most of the mound leaching probably occurred), general exposure of the basin shelf would leave a broad, shallow evaporite "pan" shoreward of the exposed mound developments. The interrelations of the basin environments would then be suited ideally for the operation of the seepage reflux dolomitization theory proposed by Adams and Rhodes (1960) to explain dolomitization in the Permian basin of West Texas. Application of this theory accounts for much of the dolomitization and anhydrite infilling of the Paradox basin mound reservoirs, especially those of the Akah and Barker Creek cycles which were deposited during the maximum development of evaporitic conditions in the basin (Fig. 2).

STRATIGRAPHIC-STRUCTURAL RELATIONS, ANETH FIELD AREA

The general structural pattern of the Aneth area is in Figure 13. Structural contours for this map are drawn using as datum the base of the black shaly dolomite overlying the Desert Creek cycle (*i.e.*, on the *top* of the Desert Creek cycle). The contours are based on well control and therefore are somewhat generalized, although well density throughout the area is sufficient to provide very adequate structural interpretation.

Important structural features of the general Aneth area are: the Bluff anticline (the southeast extension projects as a nose across the north end of the Aneth area); an east-west syncline extending east from the south side of the Recapture Creek field across the center of the Aneth mound; the faulted Desert Creek anticline; the Akah, Boundary Butte, and White Mesa anticlines; and the Ute Mountain nose which extends west across the north side of the Ismay field (Fig. 13). Aneth, the major field of the area, is lower structurally than any of the other fields and also has little or no closure. Most of the productive area of this field is in the synclinal zone between the Desert Creek and Bluff-Ute Mountain anticlinal trends. Similarly, the productive area of the Bluff field is south of the structural

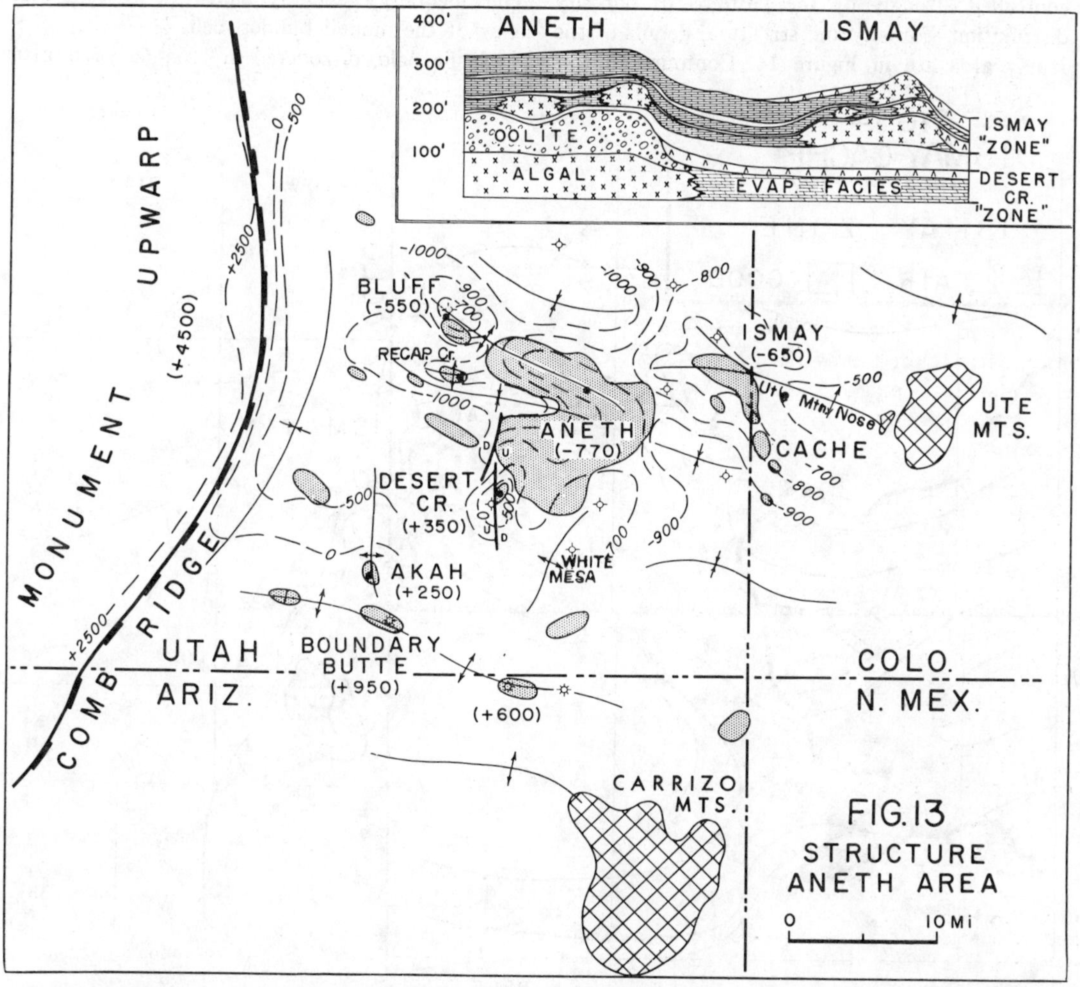

FIG. 13.—Generalized structural pattern, Aneth area. Structural contour datum is top of Desert Creek "zone" (see schematic west-east cross section, upper right). Contours (in feet) are drawn with irregular interval pattern to show general structural trends. Contour interval = 100 ft. Production from Desert Creek and Ismay cycles is shown by stippled pattern. Dry holes and marginal field discoveries (solid circles) drilled prior to Aneth discovery (solid circle) are shown.

axis, and production at the Desert Creek field is on the downthrown side of the faulted anticline. These patterns of accumulation are controlled directly by the distribution of the carbonate reservoir porosity as shown by the pattern in Figure 6.

Stratigraphic-Structural Relations, Ismay Field Area

The general structural pattern and porosity distribution in the Ismay field area are in Figure 11. This map illustrates again that most of the production is found off the structural axes and is controlled strongly by the patterns of porosity distribution. Porosity and structural details of the Ismay area are in Figure 14. Contours on this map also are based on well control and structural datum is the top of the Desert Creek cycle. The Ismay field discovery well was drilled on the axis of the Ute Mountain nose but most of the productive area is south of the nose axis where the main belt of mound development occurs. Within the Ismay cycle porosity belt, optimum mound development is discontinuous in each subcycle but tends to be localized repeatedly in the same general area, although a slight shifting of maximum thickness pattern is observed (cross section, Fig. 13). There is maximum production where all three subcycles of the Ismay cycle are developed coincidentally, generally along the eastern margins of the mound buildup belt.

Cache field, discovered in 1964 (8 years after

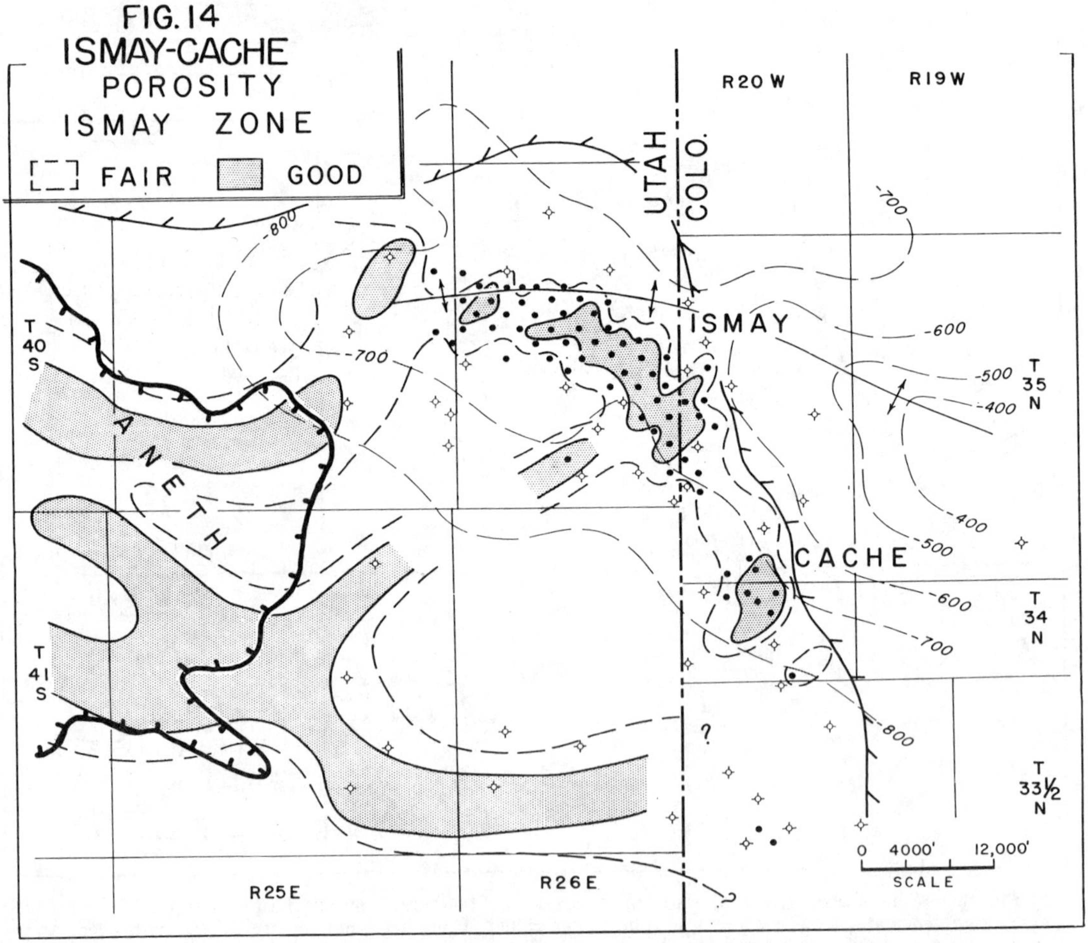

Fig. 14.—Porosity pattern, Ismay cycle, Ismay-Cache area. Structural contour datum is top of Desert Creek "zone" as on Figure 13. Contours in feet. Contour interval = 100 ft. Producing wells in Ismay-Cache area shown by solid circles. Boundary of Aneth field shown by heavy hachured line. Lower Ismay evaporite edge shown by diagonal hachured line.

the Ismay field discovery), is productive from all three Ismay subcycles in a mound buildup distinct from those of the Ismay field area, although along the same general porosity belt. This field is 50–100 ft. structurally lower than the Ismay field but contains some of the best wells of the area.

Summary

The statement can be made with considerable documentary evidence that oil accumulation in the Aneth and Ismay areas is controlled first of all by distribution patterns of individual carbonate-reservoir-mound buildup trends, together with porosity-permeability discontinuities within them. Structural influence is limited to local redistribution of earlier oil accumulations within the isolated reservoir geometry during subsequent development of the structural framework of the area. Proper location of the reservoir belt, with respect to black sapropelic shale that apparently is the petroleum source rock, is necessary, but is given secondary importance because of the widespread vertical and lateral distribution and interbedding of these rock types with the reservoir sequences.

The Aneth and Ismay fields both are primarily stratigraphic traps. The Aneth field is an isolated carbonate-mound reservoir in which the oil accumulation has undergone only minor redistribution because of later and relatively insignificant structural growth. The Ismay and Cache fields are primarily accumulations related to discontinuous porosity buildups in an algal-mound belt on the broad flank of the regional Ute Mountain structural nose. Secondary minor local shifting of the earlier accumulations may have occurred during Laramide and later structural growth.

The Aneth field provides an excellent example for illustrating the potential of many unexplored areas of the Rocky Mountains. Prior to its discovery, seven abandoned wildcat wells, in addition to the discovery wells of the marginal Desert Creek, Bluff, Recapture Creek, and Akah fields, had been drilled on surrounding structural features. This giant field, literally surrounded by dry holes and marginal wells, was discovered by drilling a well at a relatively low position on the axis of a structural nose, only after most of the more significant structural closures of the area had been drilled with only a relatively insignificant record of success.

Selected References

Adams, J. E., and M. L. Rhodes, 1960, Dolomitization by seepage refluxion: Am. Assoc. Petroleum Geologists Bull., v. 44, no. 12, p. 1912–1920.

Choquette, P. W., and J. D. Traut, 1963, Pennsylvanian carbonate reservoirs, Ismay field, Utah and Colorado, *in* Shelf carbonates of the Paradox basin: Four Corners Geol. Soc., p. 157–184.

Conner, D. C., and E. M. Wright, Jr., 1958, Bluff field, *in* Guidebook, Geology of the Paradox basin: Intermtn. Assoc. Petroleum Geologists, p. 257–261.

Elias, G. K., 1963, Habitat of Pennsylvanian algal bioherms, Four Corners area, *in* Shelf carbonates of the Paradox basin: Four Corners Geol. Soc., p. 185–203.

Intermountain Association of Petroleum Geologists, 1961, Oil and gas fields of Utah, a symposium.

Lauth, R. E., 1958, Desert Creek field, San Juan County, Utah, *in* Guidebook, Geology of the Paradox basin: Intermtn. Assoc. Petroleum Geologists, p. 275–277.

Linscott, R. O., 1958, Petrography and petrology of Ismay and Desert Creek zones, Four Corners region, *in* Guidebook, Geology of the Paradox basin: Intermtn. Assoc. Petroleum Geologists, p. 146–152.

Murray, R. C., 1960, Origin of porosity in carbonate rocks: Jour. Sed. Petrology, v. 30, no. 1, p. 59–84.

Ohlen, H. R., and L. B. McIntyre, 1965, Stratigraphy and tectonic features of Paradox basin, Four Corners area. Am. Assoc. Petroleum Geologists Bull., v. 49, no. 11, p. 2020–2040.

Parks, J. M., 1958, Plate-shaped calcareous algae in late Paleozoic rocks of Midcontinent (abs.): Geol. Soc. America Bull., v. 69, no. 12, pt. 2, p. 1627.

Peterson, J. A., 1959, Petroleum geology of the Four Corners region: Proc. 5th World Petroleum Cong., Sec. 1, Paper 27, p. 499–523.

——— and H. R. Ohlen, 1963, Pennsylvanian shelf carbonates, Paradox basin, *in* Shelf carbonates of the Paradox basin: Four Corners Geol. Soc., p. 65–79.

Picard, M. D., 1958, Subsurface structure, Aneth and adjacent areas, San Juan County, Utah, *in* Guidebook, Geology of the Paradox basin: Intermtn. Assoc. Petroleum Geologists, p. 226–230.

——— Boyd R. Brown, A. J. Loleit, and J. W. Parker, 1960, Geology of Pennsylvanian gas in Four Corners region: Am. Assoc. Petroleum Geologists Bull., v. 44, no. 9, p. 1541–1569.

Pray, L. C., and J. L. Wray, 1963, Porous algal facies (Pennsylvanian), Honaker Trail, San Juan Canyon, Utah, *in* Shelf carbonates of the Paradox basin: Four Corners Geol. Soc., p. 204–234.

Quigley, M. D., 1958, Aneth field and surrounding area, *in* Guidebook, Geology of the Paradox basin: Intermtn. Assoc. Petroleum Geologists, p. 247–253.

Robinson, B. B., 1958, Ratherford field, San Juan County, Utah, *in* Guidebook, Geology of the Paradox basin: Intermtn. Assoc. Petroleum Geologists, p. 28–286.

Wengerd, S. A., 1958, Pennsylvanian System of Four Corners region: Am. Assoc. Petroleum Geologists Bull., v. 42, no. 9, p. 2048–2106.

——— 1962, Pennsylvanian sedimentation in Paradox basin, Four Corners region, *in* Pennsylvanian System in the United States: Am. Assoc. Petroleum Geologists, p. 264–330.

Wray, J. L., and K. Konishi, 1960, Pennsylvanian and Permian codiacean algae (abs.): Geol. Soc. America Bull., v. 71, no. 12, pt. 2, p. 2006.

Reprinted from
BULLETIN OF THE AMERICAN ASSOCIATION OF PETROLEUM GEOLOGISTS
VOL. 51, NO.12 (DECEMBER, 1967), PP. 2462-2467, 1 FIG.

GEOLOGICAL NOTE

FACTORS CONTROLLING SHAPE AND INTERNAL FACIES DISTRIBUTION OF ORGANIC CARBONATE BUILDUPS[1]

ROBERT J. STANTON, JR.[2]
College Station, Texas 77843

Introduction

The geologic literature contains descriptions of many ancient and Recent "reefs," but little has been done to synthesize these data so that, when another ancient reef is discovered, the petroleum geologist can develop most efficiently the hydrocarbon reservoir within it. The purpose of this note is to consider the factors which might best serve as the independent variables needed to construct the framework for such a synthesis. The term "buildup" is used in this paper to include all those essentially organic carbonate masses (1) to which names such as "reef," "bioherm," "organic reef," "biohermal reef," and "mound" have been applied, (2) which represent predominantly in-place accumulation of largely skeleton-derived carbonate sediment, and (3) which had some topographic expression above the sea floor during growth. The term "buildup" perhaps is colloquial but is useful (1) because it is general enough to include all the forms that have been named, and (2) because it allows geologists to bypass the specific definitions and connotations of these names—a subject of secondary importance for the petroleum geologist. Stromatolitic mounds, inorganic accumulations such as oölitic bars, and deep-water coral banks such as those described by Stetson *et al.* (1962) are not considered.

This note is not concerned with the regional factors which control the distribution of buildups, but with the primary factors which determine the reservoir characteristics of individual buildups. The quality and distribution of porosity are most clearly related to external form and internal facies pattern. Therefore, a synthesis of knowledge of buildups should be in terms of the factors which determine these parameters. It is concluded that the independent variables most important in determining form and internal facies, and therefore most useful as a framework for such a synthesis, are three aspects of the physical environment: (1) the directional features of the climate, particularly the wind, which control the production of carbonate sediment and its final distribution pattern; (2) the paleogeographic setting, which determines the extent to which the directional climatic aspect can be effective; and (3) local bathymetry, which governs the final facies pattern within the buildup. The composition of the biota is less important. A rigid skeletal framework is primarily a biological response to a rough-water, high-energy environment, and its presence or absence is a relatively minor factor in the analysis of the geometry and facies of a buildup. The apparent lack of skeletal framework in many of the buildups which formed in broad shallow epicontinental seas can be evaluated best in light of the protection from rough water which was afforded by the shallowness itself. The organisms living on buildups as they were forming should be considered primarily as sediment producers.

Factors Controlling Form and Facies of Buildups

A synthesis of the diverse types of buildups in the geologic record should be made in terms of the independent variables which are most important in determining their external shape and internal facies arrangement. The factors which seem most likely to be significant are geologic age, biota, and physical environment.

[1] Manuscript received, December 16, 1966; accepted, April 10, 1967.

[2] Texas A&M University. Formerly with Shell Development Company (a Division of Shell Oil Company), Exploration and Production Research Division, Houston, Texas—(EPR Publication 473). The writer is indebted to H. A. Lowenstam, James Lee Wilson, F. J. Lucia, and L. H. Sawatsky for their stimulating discussions.

GEOLOGIC AGE

In the past, one of the main obstacles to a synthesis of existing knowledge has been the impression that buildups of different geologic ages are essentially different. Geologic age is certainly an independent variable, but it is expressed only through the biota and the physical environment; through geologic time, the biota available to produce, and perhaps to bind, carbonate sediment have changed and, at any specified locality, the physical environment has changed. Consequently, it is more fruitful to analyze buildups directly in terms of biota or physical environment than in terms of age. In addition, the fact that buildups of different ages may contain the same gross biota and be very similar in shape and internal facies emphasizes this conclusion. For example, the Mississippian mounds of the Sacramento Mountains, New Mexico (Pray, 1958), and some of the Silurian mounds of northern Indiana, are similar in shape, facies pattern, and gross biota. Both the Silurian and Mississippian mounds are circular to oval in plan view. Both consist of a central core facies of "lime mud" and flanking facies of coarser grained carbonate sediment. Bryozoa are the dominant organisms in the core facies; calcite-cement-filled voids, referred to as *Stromatactis,* also are characteristic of the mound cores. In the flank facies that appear to be contemporaneous with the core facies, crinoids are the dominant fossils.

BIOTA

The usage in geology of the term "reef" has been fairly well standardized since Lowenstam (1950) phrased it in terms of the ecologic potential of the biota to build a wave-resistant structure into turbulent shallow water and, to a certain extent, to create their own environment. As a result, in part, of the biological emphasis in this definition, it has commonly been considered that the biota are of prime importance in the study of carbonate buildups.

The composition of the biota present in a buildup must necessarily affect the facies within it, for a buildup containing massive and encrusting organisms such as the colonial corals and algae of modern reefs is far different from one containing primarily solitary and nonencrusting organisms such as crinoids, bryozoans, mollusks, and solitary corals. In addition, differences in the texture of the sediment are determined by differences in the skeletal microstructure of the various organisms that may be present on the buildup. The effects of the microstructure in determining the manner in which the carbonate sediment-producing skeleton breaks down after death of the organism are evident in ancient buildups and are comparable with those in Recent reefs described by Ginsburg (1956).

The composition of the biota controls the local depositional environment to the extent that different organisms are able in varying degrees to modify the environment around them. However, in a more general sense, biotic composition is not an independent factor but is controlled by the physical environment. The organisms living on a buildup and providing the carbonate sediment for its further growth are those adapted to the environment that exists there. The subsequent modification of the environment by the organisms is then a secondary result. This is true in spite of the clear effects that the biota may have on the environment (Ginsburg and Lowenstam, 1958). This conclusion is inherent, too, in the recognition by Lowenstam (1950) and by Textoris and Carozzi (1964) of different energy stages characterized by different biota. The Alacrán reef complex of Campeche Bank provides a localized example which demonstrates that buildup shape and facies distribution are not primarily controlled by the biota. There, the communities on the lagoonal patch reefs and on the reefs of the windward and leeward margins are similar, except for the relative scarcity of *Acropora* in the lagoonal microatolls (Kornicker and Boyd, 1962). The great differences in form and facies between the lagoonal patch reefs and the reefs of the outer margins are not due primarily to the relative abundance of this coral. Rather, all differences in form, facies pattern, and the abundance of *Acropora* probably are related primarily to the water roughness.

Lowenstam's (1950) essentially biological definition segregates within the general class of buildups those which have built into rough water and which, at least during their final stage of growth, have a rigid skeletal framework. However, unless organisms are present which possess functional morphologic characteristics indicative of rough water, one can only recognize that a buildup has grown into turbulent surface water and is presumably a wave-resistant structure from debris formed by the wave erosion to which it is resistant. The missing element of this reef definition is how rough the water must be and, correspondingly, how wave-resistant a buildup must be to be a reef. If a buildup, with or without a rigid skeletal framework, were able to grow from the sea floor upward into shallow water, it would not be classified as a reef according to this definition if, when it reached the surface, the water was not turbulent. The Silurian "reefs" of northern Illinois and Indiana grew through a quiet-water stage to a final rough-water stage (Lowenstam, 1950; Textoris and Carozzi, 1964). If they had not reached the rough-water stage because of some external cause, such as a changing environment which stopped their growth while they were still in deeper water, or if the rough-water environment ceased before they had grown into shallow water, the diagnostic capping veneer of wave-eroded debris and the organisms adapted to rough water would be lacking. Thus, they would not now be recognized as reefs according to this definition. The focus of this definition on skeletal framework and the potential to grow into shallow water, although important in paleoecology is of secondary value when considering buildups as petroleum reservoirs.

An argument for placing major emphasis on the biota when considering the formation of buildups is to point out the apparent scarcity of frame-building organisms and of buildups with demonstrable skeletal framework during some times in the geologic past, such as the late part of the Paleozoic and the early part of the Mesozoic Eras (Twenhofel, 1950). It can be argued that reefs were scarce because of the absence of frame-building organisms; and that, if frame-building organisms had been present, the buildups formed then would have been very different. In evolutionary terms, however, frame building is the adaptive response of a species as it attempts to occupy the ecologic niche where a rigid frame would be advantageous (*i.e.*, in a high-energy, rough-water environment). Because representatives of the same taxa that were important frame builders in the Devonian—hydrozoans and rugose and tabulate corals—existed during this period, the potentiality for frame building or for frame-building organisms to evolve out of these taxa also must have existed. It is as reasonable to conclude either that rough-water niches which could be occupied by existing or evolving frame builders did not exist during this time, or that the rough-water buildups which formed then with their skeletal framework have not been observed, as it is to conclude that, with taxa present having the frame-building potential, the rough-water niche was unoccupied during a span of more than 100 million years.

PHYSICAL ENVIRONMENT

Although the rigid frame and wave-resistance distinctions are extremely significant in studying buildups in terms of their ecology or paleoecology, they are not of prime importance in determining the form and facies of the buildup. The more important and more general characteristic when considering buildups as petroleum reservoirs is the formation, as a result of locally concentrated organic productivity, of a sedimentary mass rising above the surrounding sea floor. The basic role of the organisms is to produce sediment. Buildups can be compared best in terms of the physical aspects of the environment which (1) control the organic productivity and thus sediment production, and (2) shape the sediment supplied into the resulting buildup.

The aspects of the environment which appear to have the greatest effect on the development of a carbonate buildup are: (1) climate—in particular, the directional attributes expressed in the pattern of wind and storm; (2) paleogeography—the location of the buildup relative to basin, shelf, or shore; and (3) sea-floor topography adjacent to the buildup.

Climate.—The effects of wind on reef growth and sedimentation have been discussed by numerous authors in the past (*e.g.*, Krempf, 1930; Yonge, 1951; Wells, 1957). Wind, by generating waves and currents, affects the nature and circulation of the water mass surrounding the buildup, and thus controls the organic carbonate productivity and, as a result, the extent and distribution of sediment production. In general, the organic carbonate productivity is much greater on the windward than on the leeward side of modern reefs.

The energy applied to shaping a buildup is primarily in the form of wind-generated waves. The wind speed and duration and the fetch determine the amount of energy that can be transmitted by the wind to the waves. Both the prevailing winds and storm winds—which are not necessarily from the same direction—control the erosion, movement, and redeposition of the carbonate sediment. Movement of sediment around the sides of a buildup results in a characteristic horseshoe or wind-swept shape. Because of greater organic productivity and because of greater wave energy which sorts the sediment, piling up the coarser and transporting the finer clasts, coarser sediments tend to be concentrated along the windward margins of modern reefs. Coarse debris derived from the reef edge and tossed back over it forms rubble ridges and blanket- or dune-shaped deposits. Debris derived from the reef edge and deposited in front of it forms talus deposits.

Islands are more common on the windward side of modern reefs (Wiens, 1962, provides statistics for Pacific reefs). The application of this fact to ancient carbonate buildups must be emphasized, for along with subaerial exposure go the vadose processes of solution and cementation which may have significant effects on the reservoir characteristics of the sediment.

Paleogeography.—The directional climatic factors impart an asymmetry to modern buildups and are thus important in predicting the form and facies of buildups in the geologic record. The geographic setting and the topography of the sea floor adjacent to the buildup are important because they determine the effectiveness of these directional aspects of the environment.

The effect of windward barriers in limiting fetch length and thus limiting the application of wind wave energy on a buildup has been discussed in the past (*e.g.*, Cloud, 1952, p. 2140; Lowenstam, 1950, p. 466–467). What has not been emphasized sufficiently is the limiting effect of water depth alone on the amount of energy that may be imparted by wind to waves. Consideration of this effect may be extremely important in analyzing buildups which formed in broad but shallow epicontinental seas. As waves move into water of decreasing depth, energy is lost by the waves through bottom friction and percolation, and wave height, a measure of the energy, is reduced (Keulegan and Krumbein, 1950). In a shallow sea of relatively uniform depth, the height of waves that can be generated is limited by the water depth. Once waves of the maximum height have been generated, energy is lost by the waves as fast as it is imparted to them by the wind. The interrelations of fetch length, water depth, and wave height, as summarized by Bretschneider (1958) and as portrayed most graphically by him (1954, Fig. 3) are reproduced in part in Figure 1 of this note.

When one considers buildups in terms of the geographical factors of windward shoal water as well as barriers, and considers in addition that the effectiveness of winds in generating waves is limited if the fetch is limited laterally (Saville, 1954), it is evident that wind-generated waves will most effectively mold buildups which are leeward of deep water and distant from any windward land or shoal. Modern horseshoe-shaped buildups are best developed in such locations. Similar directional effects of the environment are marked also in those Upper Devonian Leduc buildups of Alberta which were in a mid-basin paleogeographic setting. The circular outline in plan view of lagoonal patch reefs and their radial symmetry of growth and of flanking deposits have been explained generally by growth in quiet deep water or in protected shallow water behind a windward barrier. The protection provided to buildups by shallow water alone, however, has not been emphasized sufficiently. For example, although hurricanes frequently cross the Bahama platform, the shallow-water depth of 2–3 fm on the broad platform limits the wind energy that can be transmitted to the waves and thus limits the power of hurricanes to erode and to transport sediment. As a result, patch reefs south of Andros Island and scattered elsewhere on the Bahama platform, which are not protected by windward shoal barriers but only by the sheltering effects of shallow water, are roughly circular and without downwind "tails" of patch-reef-derived detritus. A similar explanation may account for the numerous buildups which formed in the broad, relatively shallow Paleozoic epicontinental seas of North America and which are commonly circular to oval in plan, apparently lack a wave-resistant skeletal framework, and show little evidence of any directional aspect in the environment during growth. Differences between Silurian reefs of the Great Lakes area can also be explained in terms of the paleogeographic settings of the various reefs. The Marine reef in southern Illinois, near the edge of the Silurian Bainbridge basin, is horseshoe-shaped and concave away from the basin (Lowenstam, 1952). By contrast, the Silurian reefs in Iowa (Hinman, 1963) and in northern Indiana, far back on the shelf, are circular to oval, with flank deposits dipping radially

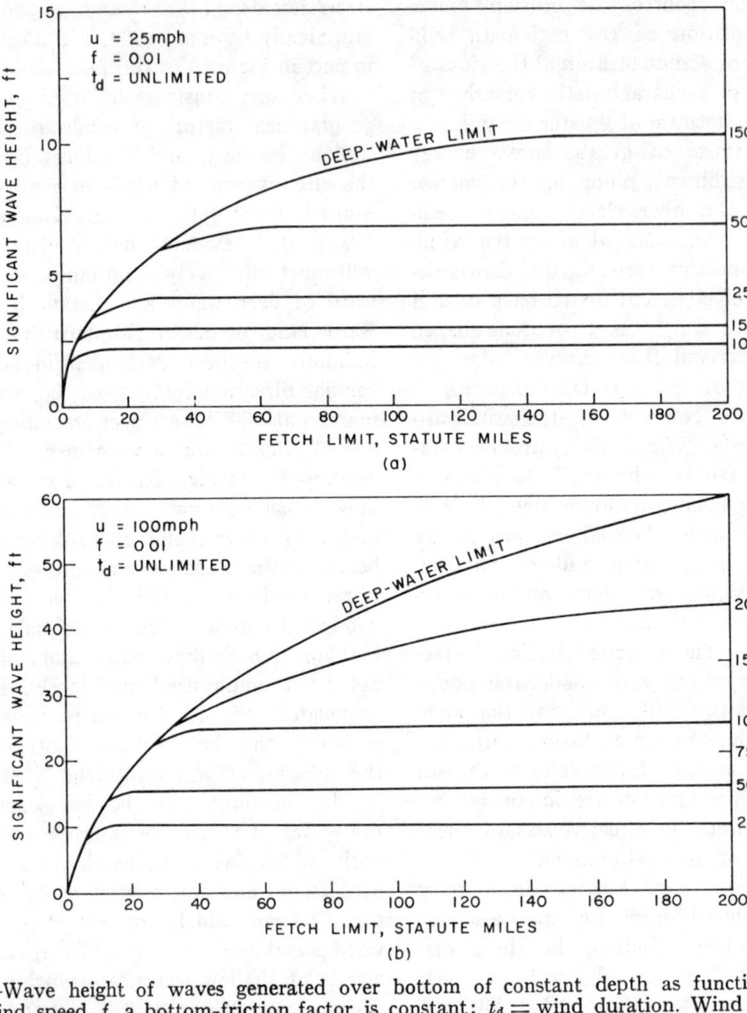

FIG. 1.—Wave height of waves generated over bottom of constant depth as function of water depth, fetch, and wind speed. *f*, a bottom-friction factor is constant; t_d = wind duration. Wind speed is 25 mph in (a), 100 mph in (b) (from Bretschneider, 1954, Fig. 3).

away and with relatively little evidence of erosion and sediment redistribution.

Local bathymetry.—The sea-floor topography adjacent to a buildup controls the effectiveness of the directional attributes of the climate in a manner similar to that of the more regional topographic features discussed in the preceding section, but probably to a lesser degree. The more marked effect of the local bathymetry is to control the pattern of deposition of the sediments flanking the buildup. For example, on the southern, stormward side of Pacific atolls, much sediment eroded from the reef edge moves down the face of the reef, gets below the effective wave base, and remains deposited as a forereef talus (Ladd *et al.*, 1950, p. 423). If however, the sea floor is relatively shallow in front of the reef, eroded material may not be able to move downward to depths below the effective wave base before being picked up by large storm waves and then deposited behind the reef. Such a situation may exist at Key Largo Dry Rocks in the Florida reef tract, where a terrace less than 30 ft deep fronts the reef. There, a forereef detritus fan is absent, and back-reef movement of reef-derived material during storms is large (Ball *et al.*, 1967).

CONCLUSIONS

Since Lowenstam (1950) emphasized the ecologic aspects of organic carbonate buildups in his study of the Silurian reefs of Illinois and Indiana, geologists have tended to maintain this viewpoint and to treat organic carbonate buildups as essentially biological entities. The value of this viewpoint is unquestioned; Ginsburg and Lowenstam (1958) have shown clearly what insights into the process of carbonate deposition can be gained by examining in detail the role of the organisms. Nevertheless, the physical environment is the prime factor in determining the nature of a buildup both by controlling the presence or absence, the growth rate, and the morphology of the organisms which produce the skeletal sediment, and by controlling the shaping of the resulting sediments into a lithologic body.

The geologist's first concern when directing the development of a reservoir in an organic carbonate buildup is to predict both the external and the internal geometry of the buildup. The thesis of this note is that geologists can do this best if, first, they are able to gauge the effects of the principal aspects of the physical environment on the sediment production, deposition and diagenesis in buildups and, second, if they can compare buildups by relating them in terms of the environmental aspects that were of prime importance during their growth. The framework within which to predict the characteristics of new buildups and within which to compare and synthesize known buildups is best constructed of (1) the directional aspects of the climate—direction of wind and storms, (2) the paleogeographic setting, and (3) the local bathymetry. The directional aspects of the climate impart an asymmetry to the shape and to the facies distribution of the buildup: horseshoe shape; greater sediment production, concentration of islands and resulting vadose processes, relatively coarser grained sediment, and potential forereef talus to windward; potential storm debris to leeward. The paleogeographic setting and local bathymetry determine the extent to which the directional climatic aspects can be effective. The protection from storm waves provided by shallow water depths alone is a particularly important factor to consider when analyzing the ancient buildups which formed in apparently unprotected settings in broad shallow epicontinental seas. The local bathymetry affects the pattern of deposition of sediment produced on the buildup.

The biota present on a buildup are determined by the physical environment and are primarily important as sources of sediment. Geologic age is important only to the extent of the changes through time in the sediment-producing and frame-building biota.

References Cited

Ball, M. M., E. A. Shinn, and K. L. Stockman, 1967, Geologic effects of Hurricane Donna in south Florida: Jour. Geology, in press.

Bretschneider, C. L., 1954, Generation of wind waves over a shallow bottom: Beach Erosion Board Tech. Memo., no. 51, 24 p.

——— 1958, Revisions in wave forecastings: deep and shallow water: Proc. 6th Conf. Coastal Eng., chap. 3, p. 30–67.

Cloud, P. E., 1952, Facies relationships of organic reefs: Am. Assoc. Petroleum Geologists Bull., v. 36, p. 2125–2149.

Ginsburg, R. N., 1956, Environmental relationships of grain size and constituent particles in some south Florida carbonate sediments: Am. Assoc. Petroleum Geologists Bull., v. 40, p. 2384–2427.

——— and H. A. Lowenstam, 1958, The influence of marine bottom communities on the depositional environment of sediments: Jour. Geology, v. 66, p. 310–318.

Hinman, E. E., 1963, Silurian bioherms of eastern Iowa: unpub. Ph.D. thesis, Iowa State Univ., 213 p.

Keulegan, G. H., and W. C. Krumbein, 1950, Stable configuration of bottom slope in a shallow sea and its bearing on geological processes: Am. Geophys. Union Trans., v. 30, p. 855–861.

Kornicker, L. S., and D. W. Boyd, 1962, Shallow water geology and environments of Alacran reef complex, Campeche Bank, Mexico: Am. Assoc. Petroleum Geologists Bull., v. 46, p. 640–673.

Krempf, A., 1930, About the shape of the coral-reefs and the system of the alternating winds: 4th Pacific Sci. Cong., v. 2A, p. 477–480.

Ladd, H. S., *et al.*, 1950, Organic growth and sedimentation on an atoll: Jour. Geology, v. 58, p. 410–425.

Lowenstam, H. A., 1950, Niagaran reefs of the Great Lakes area: Jour. Geology, v. 58, p. 430–487.

——— 1952, Some new observations on Niagaran reefs in Illinois: Illinois Acad. Sci. Trans., v. 45, p. 100–107.

Pray, L. C., 1958, Fenestrate bryozoan core facies, Mississippian bioherms, southwestern United States: Jour. Sed. Petrology, v. 28, p. 261–273.

Saville, Thorndike, 1954, The effect of fetch width on wave generation: Beach Erosion Board Tech. Memo., no. 70, 9 p.

Stetson, T. R., R. M. Pratt, and D. F. Squires, 1962, Coral banks occurring in deep water on the Blake Plateau: Am. Museum Novitates, no. 2114, p. 1–39.

Textoris, D. A., and A. V. Carozzi, 1964, Petrography and evolution of Niagaran (Silurian) reefs, Indiana: Am. Assoc. Petroleum Geologists Bull., v. 48, p. 397–426.

Twenhofel, W. H., 1950, Coral and other organic reefs in geologic column: Am. Assoc. Petroleum Geologists Bull., v. 34, p. 182–202.

Wells, J. W., 1957, Coral reefs, *in* Treatise on marine ecology and paleoecology, v. 1—Ecology: Geol. Soc. America Mem. 67, p. 609–631.

Wiens, H. J., 1962, Atoll environment and ecology: New Haven, Conn., Yale Univ. Press, 532 p.

Yonge, C. M., 1951, The form of coral reefs: Endeavor, v. 10, p. 136–144.

Reprinted from
BULLETIN OF THE AMERICAN ASSOCIATION OF PETROLEUM GEOLOGISTS
VOL. 53, NO. 11 (NOVEMBER, 1969), PP. 2314-2323, 11 FIGS.

Origin of Capitan Formation, Guadalupe Mountains, New Mexico and Texas[1]

C. W. ACHAUER[2]
Dallas, Texas 75221

Abstract The Capitan Formation originated as a linear organic bank rather than a barrier reef along the Guadalupe Mountain trend. Frame-building organisms, reef cores, and reef-derived detritus or talus are conspicuously absent in the Capitan Formation. Further, the long-standing practice of separating topographically a reef massive (core) unit above and a reef talus or breccia unit below is discounted because the same limestone lithology (mixed organic-skeletal facies) is present above and below the topographic break.

The Capitan Formation can be divided into a lower dolomite unit and an upper limestone unit. In both units, the dominant facies is characterized by much silt- and sand-size skeletal debris and to lesser extent by a wide assortment of in-place, generally scattered, invertebrate fossils. Sediment-binding algae probably were important in trapping and binding skeletal debris, thereby permitting the thick buildup of Capitan carbonate sediment.

Introduction

For the past 40 years, the Permian Capitan Formation of the Guadalupe Mountains has been considered a classic example of an ancestral barrier-reef complex separating the Northwest shelf from the Delaware basin (Fig. 1). Crandall (1929), Lloyd (1929), and Blanchard and Davis (1929) were among the first workers to advocate that the Capitan originated as a barrier reef. Later other workers perpetuated the idea of a reef origin for the Capitan (Adams and Frenzel, 1950; Hayes, 1964; King, 1948; Newell *et al.*, 1953; Tyrell, 1964). In addition, numerous guidebooks and textbooks have used the Capitan as a prime example of barrier-reef deposition. The concept that the Capitan originated as a barrier reef is firmly entrenched in the literature and in the minds of many geologists.

Baars (1964), however, did not believe that the Capitan originated as a reef. He noted the lack of reef structures and frame-building organisms in the Capitan, and suggested that the Capitan accumulated as a complex of shelf-margin carbonates in which reefs were insignificant in sedimentation.

The writer advocates that the Capitan originated as an organic bank rather than a barrier reef and summarizes the evidence indicative of an organic-bank origin. Field and laboratory studies were conducted on 18 measured Capitan sections distributed among four localities along 30 mi of the Guadalupe Mountain trend (Fig. 1). Nine of these sections in McKittrick and Slaughter Canyons provided a line of sections, averaging ¾ mi apart, completely across the strike of the Capitan Formation. Samples from all measured sections were collected at 10- or 20-ft intervals, and were slabbed, etched, and examined under a binocular microscope. Petrographic data from 469 thin sections were integrated with data from field and binocular examinations.

Capitan Subdivisions

Conventional Zonation

Conventionally, the Capitan has been differentiated into a reef breccia or talus unit and a reef limestone or massive unit. Newell *et al.* (1953, Fig. 57) and Hayes (1964, Figs. 10, 27) show the contact between these units to be transitional. The contact is placed at a prominent topographic break separating slope-forming Capitan (the reef talus or breccia unit) from scarp-forming Capitan (the reef limestone unit; Fig. 2). Genetically the reef limestone is considered to represent an accumulation of in-place reef organisms, *i.e.*, the reef core, and the reef talus or breccia unit is believed to represent carbonate detritus derived from attrition of the reef limestone at the time of reef growth and deposited in a forereef position (Newell *et al.*, 1953).

The practice of defining a boundary between reef (core) limestone and reef talus at a seemingly consistent topographic break is erroneous for three reasons.

[1] Manuscript received, December 11, 1968; accepted, January 24, 1969.

[2] Senior Research Geologist, Atlantic Richfield Company.

I am indebted to R. S. Agatston for his continued interest in this study and for critically reading the manuscript. I thank my field partners, F. E. Feldhusen and J. F. Weaver, for their perseverance and hard work in rough country.

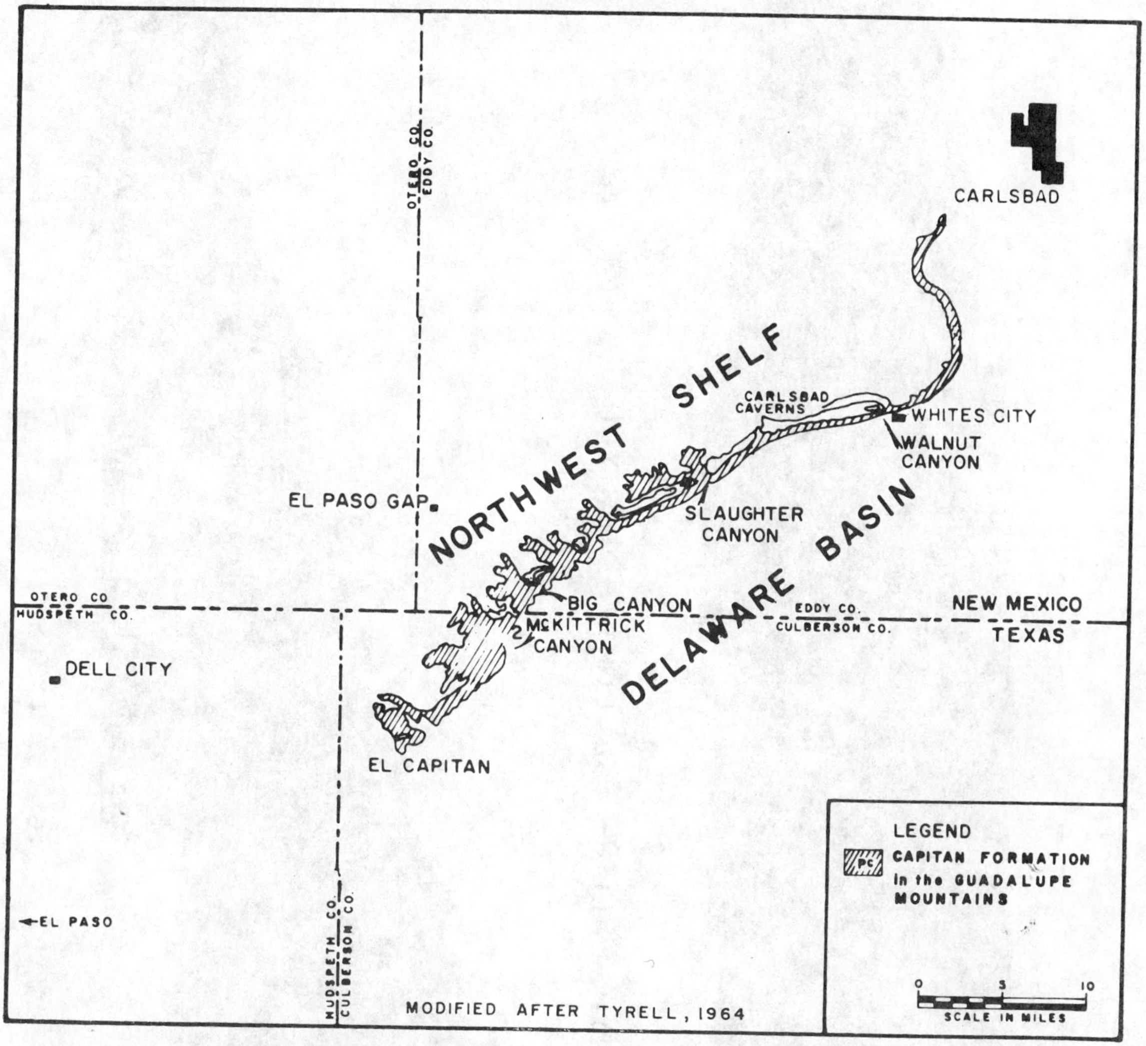

FIG. 1.—Location map of Guadalupe Mountains area, New Mexico and Texas.

1. This topographic break does not coincide with any kind of lithologic break. At all localities the same limestone lithology (skeletal calcarenite with scattered in-place organisms) has been found to lie just above and below the topographic break.

2. Prominent reef masses or cores consisting of *in situ* frame-building organisms have not been found above the topographic break.

3. Reef-derived talus or breccia has not been found below the topographic break. In essence, there is no coincidence between topography and facies.

Furthermore, the topographic break is not really consistent. The thickness of the so-called "massive" unit varies from one side of a canyon to another, suggesting that weathering and drainage effects are mostly responsible for the break.

Revised Zonation

In McKittrick and Big Canyons, the Capitan Formation can be divided into two informal lithologic units, a lower, crudely bedded, basinward-dipping dolomite unit overlain transitionally by an upper massive limestone unit (Figs. 2, 3). The combined total thickness of these two units is about 2,000 ft. Basinward the lower dolomite unit interfingers with quartz sandstone of the Bell Canyon Formation, and massive limestone of the upper limestone unit merges directly into generally distinctly bedded basinward-dipping limestone of the Lamar Member of the Bell Canyon Formation.

Shelfward, detailed studies of lithology and

NORTH WALL - SLAUGHTER CANYON

NORTH WALL BIG CANYON

field relations of the upper limestone unit indicate that its upper 300 ft interfingers abruptly with medium- to thickly bedded oölitic calcarenite or dolarenite, fusulinid calcirudite or dolarudite, and crystalline dolomite of the Seven Rivers Formation. The lower part of the upper massive limestone unit extends farther shelfward before passing abruptly into dolomite breccia (Seven Rivers Formation) thought to have originated by solution collapse of algal-laminated and "birdseye" intertidal dolomite.

The shelfward relations of the lower dolomite unit are difficult to determine because marker beds are lacking and bedding planes are poorly developed and discontinuous. If the lower dolomite shows the same shelfward relations as are seen in the upper limestone unit, it can be concluded tentatively that the lower dolomite unit flattens shelfward and passes laterally into the Seven Rivers Formation. However, more field studies are required before the relation can be resolved.

Capitan Facies

Within the upper limestone unit and the lower dolomite unit, the Capitan may be subdivided further into several distinct facies: (1) mixed organic-skeletal limestone, (2) mixed organic-skeletal dolomite, (3) sponge-algal limestone or dolomite, and (4) dolomite breccia.

Genetically, the most important facies are the widespread mixed organic-skeletal limestone, which composes much of the upper limestone unit, and its dolomitized counterpart, the mixed organic-skeletal dolomite, which makes up much of the lower dolomite unit. Facies of limited areal extent include the sponge-algal limestone and dolomite, and dolomite breccia.

Mixed Organic-Skeletal Limestone

On the outcrop and in large slabbed samples, the mixed organic-skeletal limestone facies notably lacks stratification. The facies is characterized further by fossils (in or near growth position) embedded in a matrix of silt- and/or sand-size skeletal debris (Figs. 4, 5). *In situ* fossils include bryozoans, calcareous sponges, and to lesser extent crinoids and hydrozoans. The bryozoans, particularly, are encrusted by solenoporoid algae. However, these fossils did not form a rigid organic framework, either collectively or individually.

The debris enclosing the in-place fossils is derived from a variety of organisms including crinoids, echinoids, bryozoans, brachiopods, dasyclad algae, fusulinids, foraminifers, ostracods, and many unidentifiable skeletal parts. In most samples, the amount of skeletal debris exceeds the amount of in-place fossils.

Sediment-binding algae (stromatolites) are present in many places throughout the mixed organic-skeletal limestone facies as (1) single laminae of carbonate mudstone about 0.25 mm thick, spaced several millimeters apart, and lying generally parallel with the direction of bedding, (2) domical structures consisting of very closely spaced arched laminae, and (3) weblike or anastomosing structures (Fig. 6). All these algal structures trap or enclose skeletal sand and silt. The algae probably played a major role in the stabilizing and building up of carbonate sediment in the Capitan Limestone, although a rigid organic framework was not established.

Mixed Organic-Skeletal Dolomite

The dolomite facies of the Capitan lower dolomite unit contains textures and fossils similar to those found in the Capitan upper limestone unit. Original fossils and textures were retained because replacement took place on a very small scale, involving the development of cryptocrystalline to microcrystalline dolomite. Thus, bryozoans, sponges, skeletal silt and sand, sediment-binding algae, *etc.*, can be identified in thin sections (Fig. 7). It can be concluded, from the similarity in fossils and textures, that the mixed organic-skeletal dolomite originated in approximately the same manner as the mixed organic-skeletal limestone, *i.e.*, by the production and stabilization of skeletal sand and silt along the northwest margin of the Delaware basin.

Sponge-Algal Facies

The sponge-algal facies is in the upper 150–

←

FIG. 2.—Panoramas of Big and Slaughter Canyons.

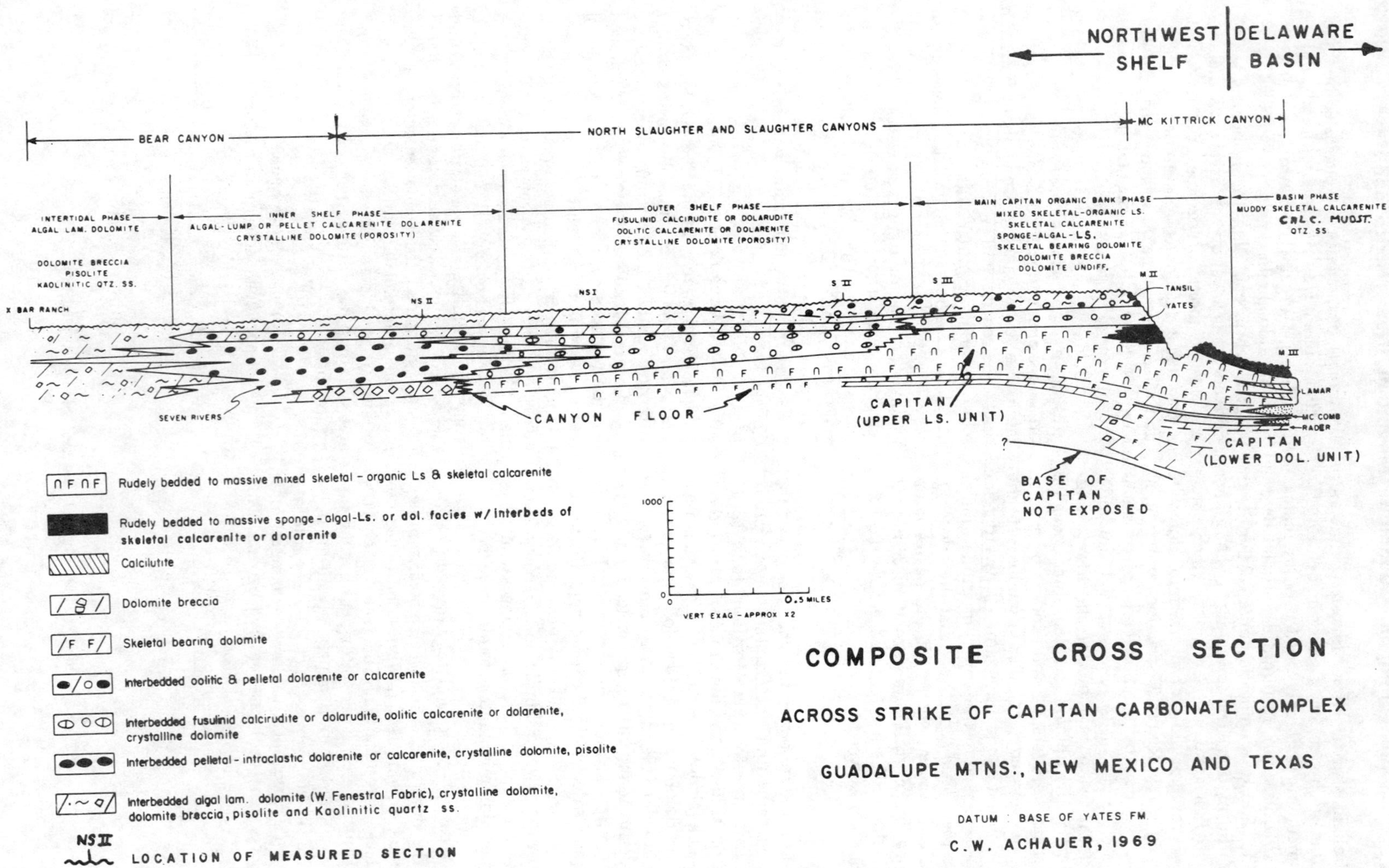

FIG. 3.—Composite cross section across strike of Capitan Formation, Guadalupe Mountains, New Mexico and Texas.

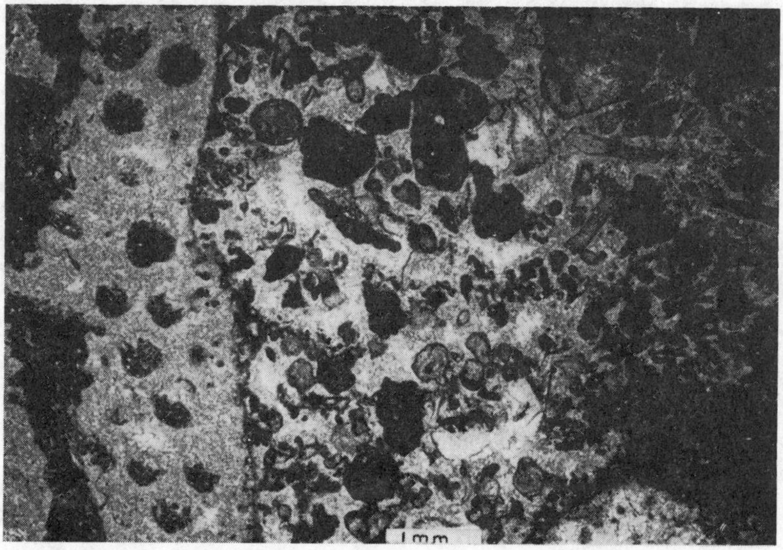

FIG. 4.—Mixed organic-skeletal limestone facies. Calcareous sponge in growth position and embedded in matrix of skeletal calcarenite.

350 ft of the Capitan Limestone. The distribution of this facies is of particular interest in that it extends for a few hundred yards on either side of the break in depositional slope of the Capitan bank (*i.e.*, the facies extends down the Capitan depositional slope to where it directly overlies the Lamar Limestone Member of the Bell Canyon Formation; Fig. 3).

The sponge-algal limestone is distinguished readily by the appearance of locally variable amounts of calcareous sponges that are supported and surrounded by laminated columnar structures of stromatolitic algae in growth position (Figs. 8, 9). Bryozoans in growth position are associated in places with the algae and calcareous sponges. Poorly defined beds of skele-

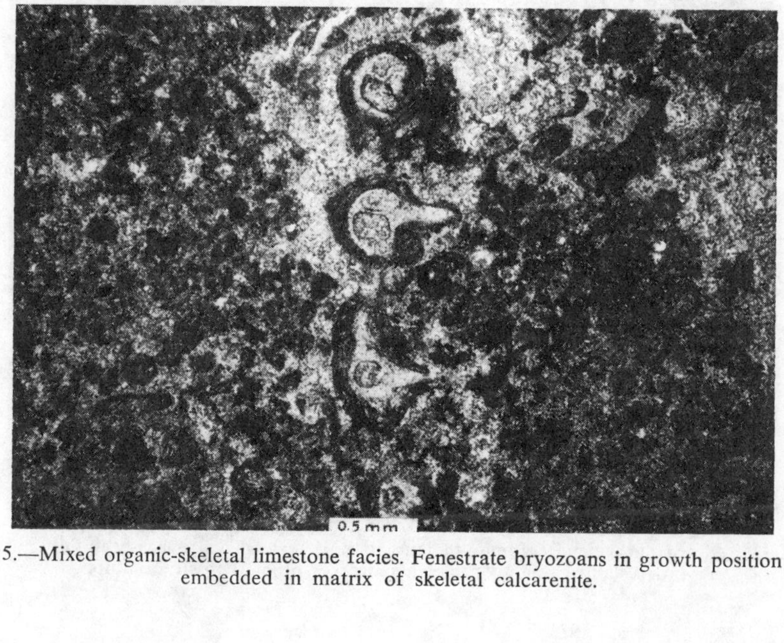

FIG. 5.—Mixed organic-skeletal limestone facies. Fenestrate bryozoans in growth position and embedded in matrix of skeletal calcarenite.

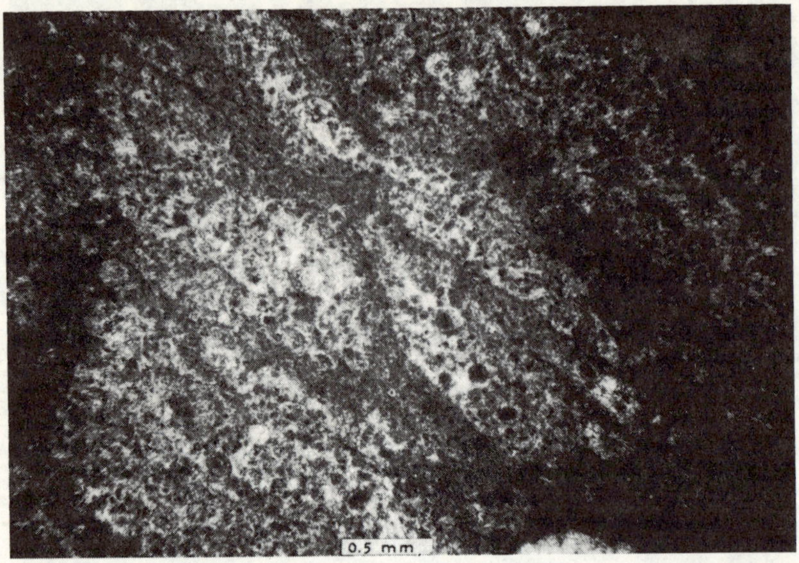

FIG. 6.—Mixed organic-skeletal limestone facies. Bifurcating structures of sediment-binding alga. Note how structures enclose and thereby trap silt- and sand-size carbonate sediment.

tal calcarenite and dolarenite also are found in the sponge-algal facies.

The organic baffle created between the calcareous sponges and the stromatolitic algae is more like a reef limestone fabric than any other facies in the Capitan Formation. Therefore, it could be concluded that the sponges and algae did build a reef. However, further inquiry into the reef-building potential of the sponge-algal facies should be made before a final conclusion is reached. For example, the amount of differential relief shown by the sponge-algal facies should be determined. Also, if the sponge-algal facies did build a reef, search should be made for flanking beds composed of reef-derived debris. Finally, the pecu-

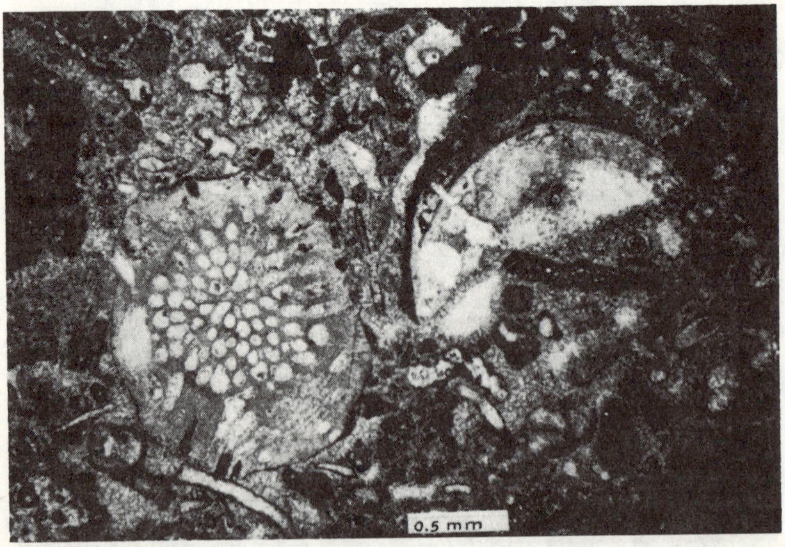

FIG. 7.—Mixed organic-skeletal dolomite facies. Much of skeletal debris has been dolomitized by cryptocrystalline dolomite, which preserves original fossil structures.

liar basinward distribution of the sponge-algal facies needs to be explained in terms of its relation to reef growth.

Dolomite Breccia

Capitan breccia appears to be concentrated near the margin of the Northwest shelf. Most is limited to the lower dolomite unit where brecciated zones 10 to nearly 40 ft thick have been observed. These zones are interbedded fairly regularly, but inconspicuously, with the mixed organic-skeletal dolomite facies.

The Capitan breccia has textures and fossil parts identical to those in the Capitan mixed organic-skeletal dolomite facies, which indicate the source of the dolomite clasts. However, there is question as to how these clasts were derived. Some breccia clasts can be fitted together along fracture planes, and thus indicate brecciation as a result of tectonic movement or perhaps solution collapse (Fig. 10). Other breccia clasts cannot be fitted together, and instead lie at various orientations in a matrix of very fine to coarse dolomite clasts (Fig. 11). In this case, it is not clear whether the breccia results from deposition, tectonic movement, or solution collapse. There are no examples in which it can be established that the breccia was derived from a reef core or reef limestone.

Origin

The determination of the origin of the Capitan Formation revolves about an assessment of the ecologic potential of organisms responsible for the Capitan buildup. In making this assessment, I follow the guidelines laid down by Lowenstam (1950) and Nelson *et al.* (1962). These workers defined a reef as a skeletal deposit formed by organisms having the ecologic potential to erect a wave-resistant structure. Nelson *et al.* (1962) further defined a bank as a skeletal deposit formed by organisms which do not have the ecologic potential to erect a rigid wave-resistant structure.

An evaluation of the ecologic potential of fossils in the Capitan Formation rests on the following considerations.

1. Most of the Capitan lacks reef cores or large masses of colonial frame-building organisms.
2. The Capitan lacks stratified deposits of bioclastic debris which can be shown to have been derived from reef cores.
3. Primarily the Capitan consists of silt- and sand-size skeletal debris derived from many kinds of organisms that thrived along the edge of the Northwest shelf.
4. Studded throughout the skeletal sand and silt are fossils in growth position. Of these, the predominant forms are calcareous sponges and bryozoans. These organisms may have been capable of forming, and actually may have formed, sediment baffles in the Capitan; however, they did not build a wave-resistant structure in the Capitan, nor probably did they have the ecologic potential to do so.
5. Sediment-binding algae are ubiquitous in the Capitan and, with other organisms, probably played a major role in trapping skeletal sediment and in permitting the thick buildup of Capitan organic-skeletal limestone facies.

Fig. 8.—Sponge-algal limestone facies. Weathered float rock accentuates baffle of calcareous sponges in growth position.

From these considerations, it is concluded that fossils in the Capitan did not have the ecologic potential to build a wave-resistant structure. The reason for the Capitan buildup is that the Capitan originated as an organic bank upon which much skeletal sediment was produced, deposited, and stabilized.

Conclusions

A consideration of lithologic, paleontologic, and stratigraphic data indicates that most of the Capitan Formation did not originate as a barrier reef. Instead, the evidence dictates that the Capitan formed as an organic bank.

Nothing seems to be gained by using the Capitan as a model for ancestral barrier-reef deposition. It would be better to typify the Capitan as a model for organic-bank deposition. It would also help to improve our understanding of Late Permian sedimentation and environments in the Delaware basin area.

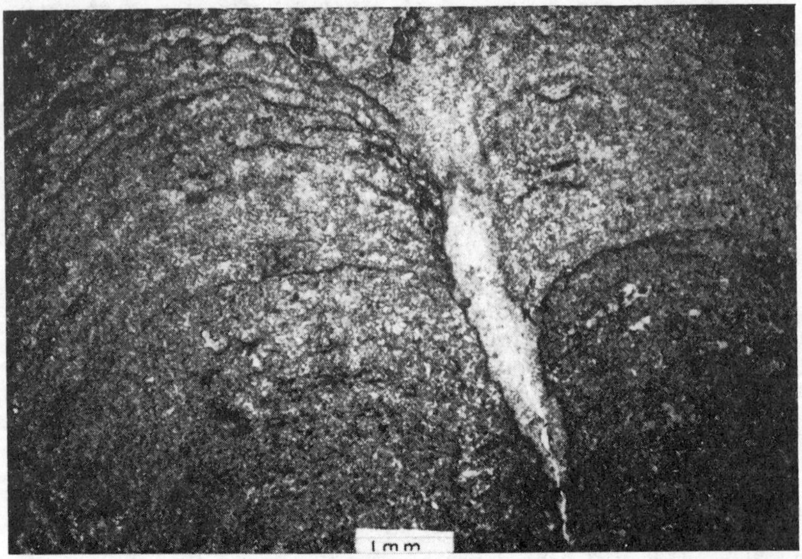

FIG. 9.—Sponge-algal limestone facies. Photomicrograph shows two columnar structures of stromatolitic alga. This alga is in close association with calcareous sponges in many places.

Unresolved Problems

The determination of the amount of original depositional dip at the front of the Capitan bank is necessary to an assessment of water depth along the bank slope and in the Delaware basin. At present, the determination of the original depositional dip is difficult. Dip readings ranging between 10° and 25° are recorded most commonly for basinward-dipping units of the Capitan lower dolomite unit and the upper limestone unit. However, locally, dips as great as 45° have been observed within calcarenite and dolarenite units that overlie the sponge-algal facies in Big and McKittrick Canyons. Just how much of the dip can be attributed to original depositional slope and how

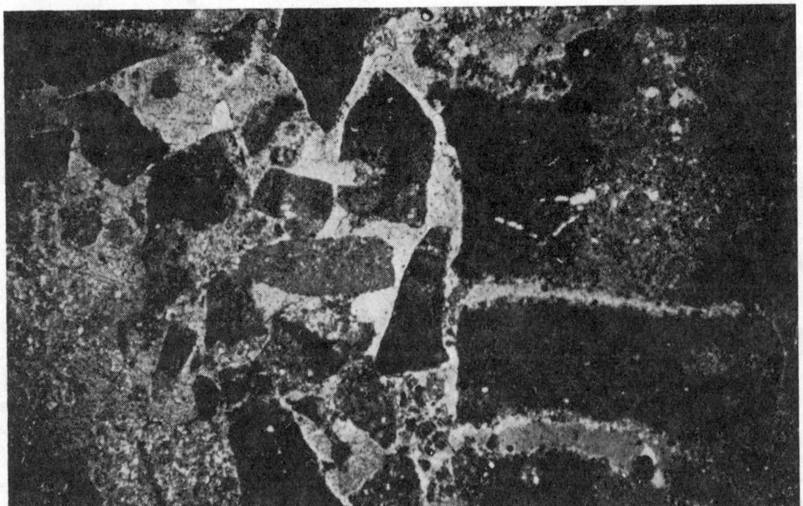

FIG. 10.—Dolomite breccia. Note that some clasts can be fitted together, and thus suggest that fracturing or solution collapse may be responsible for brecciation.

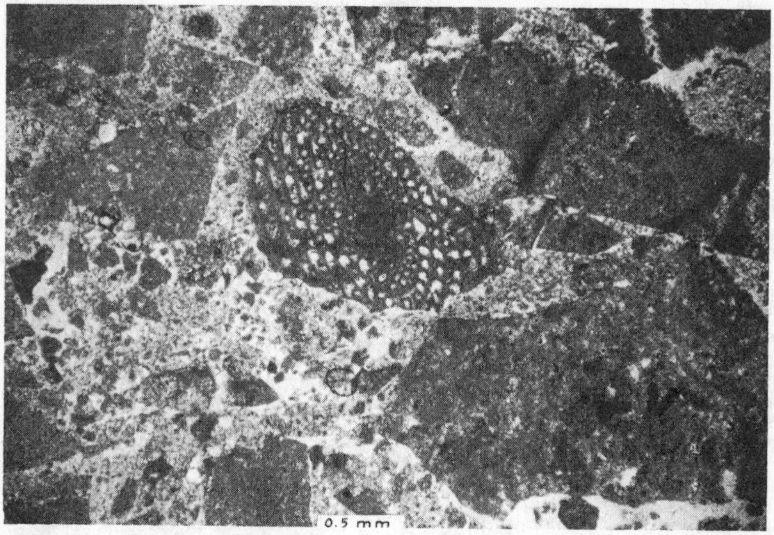

FIG. 11.—Dolomite breccia. Large clasts are variously oriented in matrix of sand-size clasts. Origin of breccia is uncertain.

much can be attributed to downwarping and/or compaction is not clear.

Another perplexing problem is the explanation of why a shallow-water facies, such as the sponge-algal limestone, directly overlies the Lamar Member at the mouths of McKittrick and Big Canyons. Basinward, the lower contact of the sponge-algal facies drops several hundred feet topographically from its position in the Capitan bank to its position at the mouths of McKittrick and Big Canyons. In short, the sponge-algal facies actually extends down the Capitan depositional slope to overlie limestone that accumulated in a forebank position. Whether the basinward extension of the sponge-algal facies reflects a significant drop in sea level after the deposition of the Capitan bank or whether some other explanation is required is a matter for further study.

References Cited

Adams, J. E., and H. N. Frenzel, 1950, Capitan barrier reef, New Mexico and Texas: Jour. Geology, v. 58, no. 4, p. 289–312.

Baars, D. L., 1964, Modern carbonate sediments as a guide to old limestones: World Oil, v. 158, no. 5, p. 95–100.

Blanchard, W. G., Jr., and M. J. Davis, 1929, Permian stratigraphy and structure of parts of southeastern New Mexico and southwestern Texas: Am. Assoc. Petroleum Geologists Bull., v. 13, no. 8, p. 957–995.

Crandall, K. H., 1929, Permian stratigraphy of southeastern New Mexico and adjacent parts of western Texas: Am. Assoc. Petroleum Geologists Bull., v. 13, no. 8, p. 927–944.

Hayes, P. T., 1964, Geology of the Guadalupe Mountains, New Mexico: U.S. Geol. Survey Prof. Paper 446, 69 p.

King, P. B., 1948, Geology of the southern Guadalupe Mountains, Texas: U.S. Geol. Survey Prof. Paper 215, 183 p.

Lloyd, E. R., 1929, Capitan limestone and associated formations in New Mexico: Am. Assoc. Petroleum Geologists Bull., v. 13, no. 6, p. 645–658.

Lowenstam, H. A., 1950, Niagaran reefs of the Great Lakes area: Jour. Geology, v. 58, no. 4, p. 430–487.

Nelson, H. F., C. W. Brown, and J. H. Brineman, 1962, Skeletal limestone classification, *in* Classification of carbonate rocks: Am. Assoc. Petroleum Geologists Mem. 1, p. 224–252.

Newell, N. D., *et al.*, 1953, The Permian reef complex of the Guadalupe Mountains region, Texas and New Mexico: San Francisco, W. H. Freeman, 236 p.

Tyrell, W. W., Jr., 1964, Petrology and stratigraphy of near reef Tansill-Lamar strata, Guadalupe Mountains, Texas and New Mexico: Roswell Geol. Soc. Guidebook, p. 66–75.

Reprinted from
BULLETIN OF THE AMERICAN ASSOCIATION OF PETROLEUM GEOLOGISTS
VOL. 54, NO. 10 (OCTOBER, 1970), PP. 1931-1932, 1 FIG.

Stratigraphic Reefs Versus Ecologic Reefs[1]

R. J. DUNHAM[2]
Houston, Texas 77001

Abstract Thick laterally restricted masses of pure or largely pure carbonate rock long have been called "reefs." Such masses as the Capitan reef are here termed "stratigraphic reefs," in contrast to organically bound "ecologic reefs." Stratigraphic reefs differ from ecologic reefs in that their binding may be either organic or inorganic. The two types of binding are detected easily in transported debris.

Thick laterally restricted masses of pure or largely pure carbonate rock are attractive as potential reservoir rock. Such masses long have been called "reefs" by stratigraphers and other geologists; they therefore are here termed "stratigraphic reefs"[3]. Since the wide acceptance of Lowenstam's (1950) ecologic redefinition, the unqualified word "reef" has come to have a narrow meaning. Lowenstam said that a reef, in terms of ecologic principles, is the product of actively building and *sediment-binding* organisms, which have the ability to erect rigid wave-resistant topographic structures. It is essential in an "ecologic reef" that organisms provide the binding.

Taken as concepts, the ecologic reef and the stratigraphic reef are basically different. The stratigraphic reef is an objective concept concerning three-dimensional geometric masses, visible from measured sections and mechanical logs. The ecologic reef is a subjective concept concerning inferred interactions of organisms and topographic evolution. One involves geometry, the other topography. The stages by which a stratigraphic reef is formed may include ecologic reef stages, but not necessarily (Fig. 1). In fact, island stages may make as great an imprint on stratigraphic reefs as do ecologic reef stages or other marine shoal stages (Dunham, 1969a).

Many stratigraphic reefs are bound to a greater or lesser degree during their growth. Their binding may be organic, as a result of ecologic reef stages, or it also may be inorganic due to diagenesis during island stages or possibly during submarine shoal stages, a conclusion drawn long ago by Bergenback and Terriere (1953). The problem of whether a geologic reef owes its binding to organisms arises whenever one hears the post-1950 question, "Was this particular reef *really* a reef?"

Examining talus helps answer the question. The presence of talus implies binding. Organic binding and inorganic binding are structurally different. In organic binding, some particles are bound but others are not—corals and encrusting algae in modern reefs are joined one to the other, but the sand grains trapped in interstices and pockets in the bound mass are themselves loose. In inorganic binding, all the particles in the bound mass are joined together—cementation is pervasive. Blocks broken from a reef and rolled downslope thus carry clues about binding. If the broken edges of a talus block consist partly or wholly of particulate sediment, the block and the reef were cemented instead of organically bound, because particles trapped in pockets between binding organisms would have fallen out.

Use of this criterion on the surface Capitan reef and on the subsurface Scurry reef indicates that the binding was wholly or largely inorganic, which accords with other evidence (Bergenback and Terriere, 1953; Dunham, 1969b; Stafford, 1959). The Capitan reef and the Scurry reef thus are examples of a large class of stratigraphic reefs that are not ecologic reefs, not "*really* reefs." Organisms provided

[1] Manuscript received, November 28, 1969; revised and accepted, March 26, 1970. Shell Development Company Exploration and Production Research Publication No. 519.

[2] Shell Development Company (a division of Shell Oil Company), Exploration and Production Research Center.

[3] Stratigraphic reefs were called "geologists's reefs" and "geologic reefs" (Dunham, 1969b) prior to March 1970, when Philip H. Heckel, from his work on classifying reefs or buildups in detail for the Calgary SEPM symposium, kindly pointed out in a letter that my geologic reefs, in fact, were stratigraphic entities deserving of the name.

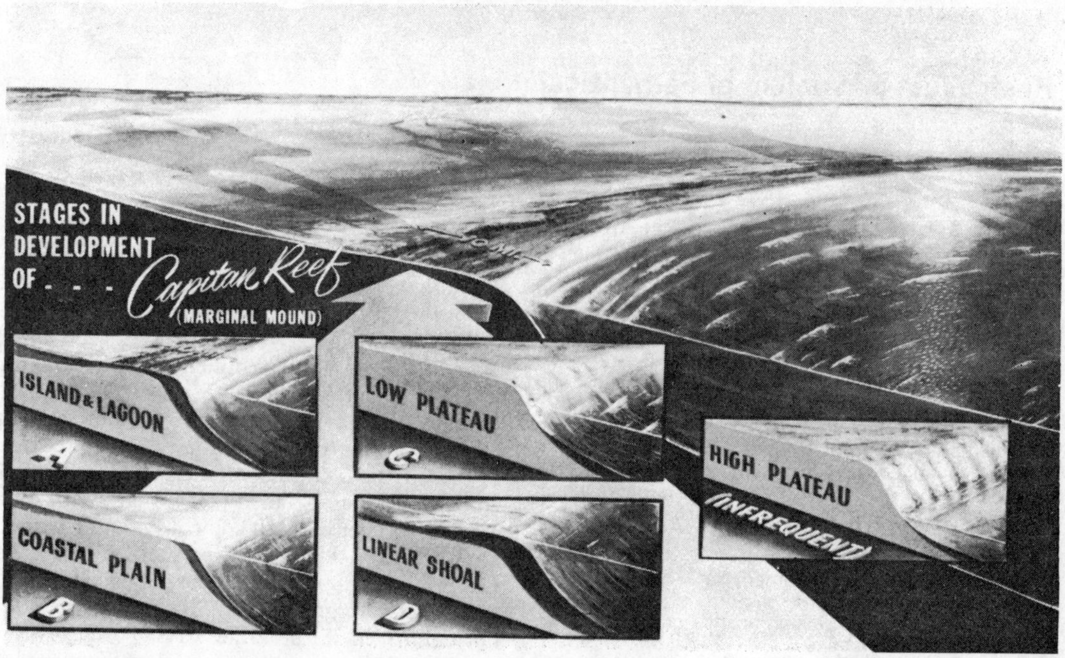

FIG. 1.—Capitan geologic reef repeatedly passed through various topographic stages.

their skeletal debris, their bulk; but organisms did not provide their rigid framework (except perhaps locally, and incidentally).

Some workers, including myself, have tried to resolve the problem of the reef concept by practically striking the word "reef" from their vocabularies, using "buildup," "margin," "mound," and such terms as substitutes. This is unsatisfactory evasion; I now recommend that we retain the word, but qualify it as needed.

REFERENCES CITED

Bergenback, R. E., and R. T. Terriere, 1953, Petrography and petrology of Scurry reef, Scurry County, Texas: Am. Assoc. Petroleum Geologists Bull., v. 37, p. 1014–1029.

Dunham, R. J., 1969a, Early vadose silt in Townsend mound (reef), New Mexico, *in* G. M. Friedman, ed., Depositional environments in carbonate rocks: Soc. Econ. Paleontologists and Mineralogists Spec. Pub. 14, p. 139–181.

——— 1969b, Vadose pisolite in the Capitan reef (Permian), New Mexico and Texas, *in* G. M. Friedman, ed., Depositional environments in carbonate rocks: Soc. Econ. Paleontologists and Mineralogists Spec. Pub. 14, p. 182–191.

Lowenstam, H. A., 1950, Niagaran reefs of the Great Lakes area: Jour. Geology, v. 58, p. 430–487.

Stafford, P. T., 1959, Geology of part of the Horseshoe atoll in Scurry and Kent Counties, Texas: U.S. Geol. Survey Prof. Paper 315-A, 20 p.

Reprinted from
BULLETIN OF THE AMERICAN ASSOCIATION OF PETROLEUM GEOLOGISTS
VOL. 57, NO. 6 (JUNE, 1973), PP. 1100-1116, 4 FIGS.

Reefs: Just a Problem of Semantics?[1]

C. J. R. BRAITHWAITE[2]
Dundee, Scotland, United Kingdom

Abstract The modern reef is both a morphologic and an ecologic concept. It is a structure which rises more or less abruptly from the sea floor to within a short distance of the water surface. This position and the topographic isolation of the structure are the primary ecologic controls for organisms living on the reef. Thus, rapid changes in depth influence water circulation and chemistry, the availability of light, and the movement of sediment.

The reef biota includes both sessile frame-building organisms and vagrant animals, zonally arranged in relation to an ecologic gradient determined by the environment. It is, however, the frame-builders which are regarded as definitive of "reef" structure.

Many ancient facies assemblies which have been described as reefs differ from such a model. The major areas of subjective argument lie in the interpretation of the depositional environments of the sediments and of the paleoecologic significance of particular organisms and associations.

There is a general lack of knowledge of the internal features of modern structures, and particularly of the influence of their recent growth history upon internal relations. This also contributes to the lack of precision surrounding reef definition.

An analysis of the criteria available for the recognition of ancient reefs shows that other structures are present in the geologic column which pass unnoticed beneath the blanket title of "reef." It is the aim of this paper to differentiate between these bodies and reefs in the modern sense. Their proper description carries both interpretative and economic benefits.

Introduction

For many authors, the recognition of reefs in the geologic column has presented no problem, but it seems fair to say that this has been less often a result of their clarity of thought than of their use of some quite personal definition. However, as it is part of the business of the scientist to communicate he should be sure that his audience has the same understanding of what he is saying as he does. The purpose of this paper is to define "reefs" and to suggest ways of recognizing them. It will first review and consider briefly the ecology of modern reefs (about which there is a general, though not complete, concensus concerning recognition) and discuss the contribution of ecologic factors to reef structure and growth. Second, it will examine some interpretations of reef paleoecology and, in the light of these, consider how fossil reefs can be recognized. This approach is justified on the grounds that, by the use of the term "reef," one is making a comparison with modern structures, as these were recognized first. To argue that this is not so is to say that some quite different structures found in fossil environments have for some reason been called "reefs." What the term then means is difficult to say.

The sedimentology of clastic rocks has developed to a point where sedimentary structures and associations often are quite well understood in terms of the physics of their formation (*e.g.*, Allen, 1971). Environmental interpretation thus can be based on a firm theoretical footing and on comparison with detailed facies analyses of modern environments. With some exceptions carbonate workers have not followed such lines of investigation. This is partly because the distribution of the skeletal fragments forming the sediments is not controlled in the same ways by the physical conditions. Grains originate within the transport system and their sizes, measured in bulk, are not a simple reflection of the speed or duration of the current. The lack of physically orientated studies also is a product of the interests of the workers who have been, for the most part, paleontologists only recently concerned with ecology, or stratigraphers. Studies equivalent to those of Jopling and Walker (1968) are not available for reef systems, and there is a regrettable tendency to interpret these environments less stringently. One major difficulty is that, by saying that a structure is a reef, an author immediately has a familiar model to which he can relate sediments. Without this model he has the more difficult task of finding suitable descriptions and a convenient analogy.

Any primary description which uses terms such as "reef limestone" or "intrareef detritus" (*e.g.*, in Terry and Williams, 1969) as descriptive terms seems to the writer to be lacking in scientific method. It is prejudging an interpretation which

[1]Manuscript received, May 1, 1972; accepted, December 8, 1972.

[2]Department of Geology, The University, Dundee.

The field work providing the experience on which this paper is based was financed by grants from the Natural Environment Research Council and the Carnegie Trust for the Universities of Scotland, whose support is gratefully acknowledged.

Discussion with colleagues has extended over the past 10 years, and it would be difficult to name all of the individuals who have been of help. Special mention should be made, however, of J. D. Taylor and B. R. Rosen, who have suffered more than most.

may later be placed on a system and using this as a substitute for a description of the specimen as, for example, a boundstone, grainstone (Dunham, 1962), biomicrite, or biolithite (Folk, 1962). Only by the use of some nongenetic term can the facts of observation be understood and stand apart from the delusions of interpretation.

Some aspects of this discussion are not new, having appeared in the literature as early as 40 years ago, but they seem worth reviewing in the light of recent contributions. The necessity for such a reappraisal stems from a lack of acceptance of the value of some of this earlier work, as there is still no generally agreed definition and usage of the term "reef." It can be argued that this has no more than a casual semantic significance, but the processes of deposition closely control the porosity of a sediment and more accurate descriptions of sequences carry economic benefits.

The literature concerning reefs and reef problems is extensive and it has been simpler to note earlier reviews rather than repeat them. Similarly, the space devoted to individual contributions has been limited, generally to asking two questions: "What is a reef?" and "Is this one?"

These questions should be asked wherever the word "reef" is used; at present, different authors are likely to give different answers.

Recent Reef Ecology and Structure

Most work on modern reefs has been by biologists and is thus largely from a biologic point of view. The investigations which could be called geologic have been concerned either with ecologic or sedimentologic features or with overall morphology. In the latter area little improvement has been made on Darwin's (1890) original classification of fringing reefs, barrier reefs, and atolls, although other minor types such as platform reefs (Odum and Odum, 1955) and reef knolls, have been added.

Darwin's (1890) ideas on the ways in which the structures grow are still popularly accepted, because they form an elegant and convenient model. Darwin's views on the mechanism of construction of reefs and atolls also remain unchallenged for certain specific areas, although they have been modified and augmented by a great variety of alternative explanations of which those of Daly (1915, 1948) are perhaps the best known. Vaughan (1919), Davis (1928), and Gardiner (1931) published extensive notes on early theories of coral reef formation. Davis (1928) is notably useful both for the way in which earlier work is reviewed and for a particularly comprehensive reference list.

As a first premise, it might be said that however one approaches a reef, whether it is to consider the ecology of the associated organisms, or simply to examine its morphology, the structure is recognized by its topographic expression. This is normally large enough to bring about variations in the surrounding physical environment. In most cases the reef is characterized by growth of corals at or close to sea level, and this position, with its consequent steep gradients of environmental change, is the root cause underlying the biotic zonation which is almost universally associated with these structures and has come to be one of the criteria for their geologic recognition.

It is convenient for the moment to overlook the fact that a reef has had any history and to consider it simply as a rocky intertidal and subtidal platform. This approach is implicit in Doty (1957) and has been advocated by Taylor (1968). The patterns and positions which various groups of organisms take on such a surface have been extensively described, sometimes in comprehensive reports such as Maxwell's (1968) Great Barrier Reef Atlas, but more often by the gradual accumulation of contributions of individual workers. Maxwell's work is in part the result of such accumulations. Reviews of modern reefs are available in Wells (1957), Milliman (1965), Stoddart (1969a), and Newell (1971). The following list of papers is a sample of those papers to which the writer has referred, that describe reefs throughout the tropics.

Bahamas: Newell and Rigby (1957); Storr (1964); Milliman (1967)
Barbados: Lewis (1960, 1965)
Bermuda: Garrett *et al.* (1971)
Bikini: Tracey *et al.* (1948); Emery *et al.* (1954)
Brazil: Laborel (1969)
Comoros: Guilcher (1971)
East Indies: Umbgrove (1939, 1947)
Fiji: Davis (1928); Gardiner (1931); Ladd and Hoffmeister (1945)
Florida: Shinn (1963); Jones (1963)
Funafuti: Finck (1904)
Gilbert Islands: Nugent (1946)
Great Barrier: Stephenson *et al.* (1931); Manton and Stephenson (1935); Maxwell (1968)
Hawaii: Edmondson (1928)
Jamaica: Goreau (1959, 1961)
Madagascar: Picard (1967); Thomassin (1969); Pichon (1971)
Red Sea: Mergner (1971); Sheer (1971)
Sahul Shelf: Teichert and Fairbridge (1948); Fairbridge (1950)
Saipan: Cloud (1959)
Samoa: Mayor (1924)
Seychelles: Taylor (1968); Rosen (1971); Braithwaite (1971)
Solomon Islands: Morton and Challis (1969); Stoddart (1969b)
Tuamotos: Doty and Morrison (1954)
Yucatan: Logan (1969)

Many of the characteristics which the structures described in these papers share can be demonstrated by reference to one example, the reefs of the Seychelles. These are not chosen because they represent some hypothetical average, but simply to use descriptions from my own experience (see Braithwaite, 1971; Braithwaite and Talbot, 1972). They differ in detail from some other reefs (the sediments, for example, include a contribution from terrigenous sources), but their general characteristics seem typical. The ecology of the corals has been considered by Rosen (1971) and that of molluscan and other invertebrate communities by Taylor (1968). Lewis (1969) has described some aspects of sedimentology.

The reefs in this area are shelflike extensions from the shore (fringing reefs) and environments are divisible into reef-edge, backreef, and basinal (*i.e.,* flat floor in front of the reef) areas. The elevation of the structures above the basinal floor lies, depending upon situation, in the range 10-15 m. This is not typical of reefs in general, where the reef front commonly terminates at greater depths.

The shoreward margin may be a beach facing areas of unconsolidated recent sediment known as "plateaux," or steep granitic cliffs. Between the margin and the reef edge is a shallow platform with water seldom exceeding 2 m depth and with wide areas emersed at low tide. This platform may be up to 500 m in width in normal circumstances, but is locally up to 2 km wide. Although waters are renewed at high tide, there are areas with poor circulation which may reach extremes of temperature (40°C) and salinity (reductions by rainfall and stream dilution or concentration by evaporation). The mean figures for open waters are 29°C and 35°/₀₀. The reef flat (backreef) is divided into three separate zones, a near-shore zone of rippled sand, an intermediate zone characterized by marine angiosperms, and an outer zone of sand and cobble ridges. Each of these is typified by particular groups of organisms, sediments, and structures. The general principles of zonation are discussed by Morton and Challis (1969), but the ideas are not new, having been used by Stephenson *et al.* (1931), Tracey *et al.* (1948), and others.

In the backreef environments, sediments are loose and unconsolidated. Some areas are dominated by marine angiosperms, whose binding and trapping activities have been well described by Scoffin (1970). As a body, the sediments commonly lack internal structures other than those produced by the activities of burrowing mollusks and crustaceans.

Scattered corals lying free on the sediment surface increase in number and diversity toward the seaward edge. It is significant that, in waters of similar depth and environmental energy, some are massive, rounded forms (*Favia* spp., *Galaxaea, etc.*) whereas others are delicately branched (*Acropora* spp.). This observation is not unique, but is difficult to reconcile with some paleoecologic analyses.

Directly behind the edge is a zone of cobbles, bound together in varying degrees by calcareous algae, which seems broadly comparable to boulder ramparts described in Pacific areas. This zone appears (Braithwaite, 1971) to represent addition to a surface lying somewhat below present sea level. The algae are frame-builders in the sense that they encrust and bind together existing debris to form an open cavernous mesh. This becomes progressively occluded on the seaward margin by algae growth within the porosity and by percolation of finer surface-derived sediment.

The edge zone is characterized by organisms which are frame-builders by virtue of their own growth, the contribution from loose sediment being generally small. The presence of frame-building organisms has long been established as one of the most widely accepted indicators of "reef" environments. In the Seychelles a series of assemblages can be recognized which represent a gradient of decreasing energy. The structural contribution of these elements is discussed elsewhere (Braithwaite, 1971), and the parallel ecological zonation is described by Rosen (1971).

There is no well defined "*Lithothamnion* zone" or the massive boulder ramparts common on some Pacific reefs. The sloping seaward faces of windward reefs in other areas are characterized by structures usually described as "spurs and grooves." There has been a continuing argument as to whether these are constructional forms produced by organic growth (*cf.* Goreau, 1959) or are erosional features on which coral growth is relatively unimportant (Cloud, 1959). These features are in general only poorly developed on Seychelles reefs, but structures which appear to be comparable are found at the southern end of Mahe, the main island. There they clearly are produced by erosion of a rocky platform which bears little coral growth (Braithwaite, 1971). The truth is probably that both views are correct in particular circumstances.

Coral morphology has been used as an important environmental indicator by some workers. Observations on modern corals suggest that their views may have been oversimplified. In the Seychelles, the following variations have been noted. Where wave action is excessive no corals can grow, but robust branching forms (*Pocillopora* and the alcyonarian *Millepora*) are among the first to appear. The significant point is that, among reef-front associations in the Seychelles, branching *Acropora* species are characteristic of areas of water movement and massive rounded forms such as *Porites* and *Platygyra* characterize quiet waters.

Branched forms, by producing a more open, porous, structure may be energy absorbing and may survive in turbulent areas even though they possess less mechanical strength.

However, one of the major controls over distribution of forms is the availability of food and oxygen. Beyond the zone in which direct wave impact limits growth, conditions of good water circulation are able to support corals with high polyp densities, such as members of the *Acropora* assemblage (Rosen, 1971). These are generally fast-growing genera and some almost monospecific stands of *Acropora formosa* are seen locally where growth has tended to smother slower growing forms. In conditions of poor circulation only more tolerant corals, such as members of the *Porites* assemblage, can survive. They are massive colonies characterized by low polyp densities per unit volume and slow growth rates in terms of speed of ground cover. Waters of intermediate character may contain mixed assemblages where the two groups are able to compete on almost equal footing.

In addition to control by water movement and the supply of nutrients, most hermatypic corals are controlled in some way by light intensities. There is thus a vertical distribution with the *Porites* assemblage appearing at depth below the *Acropora* assemblage, irrespective of any lateral changes.

Fragile sheetlike colonies (*e.g., Pachyseris*) which are strongly light-oriented occur at depth toward the limits of coral growth.

The interaction of mechanical water movements, availability of nutrients, and the distribution of light, in addition to adverse effects resulting from variation in salinity, temperature, and suspended sediment distribution, is effective in controlling colony morphology. Because many of these variables cannot be assessed in fossil systems, it is only with considerable caution that one may approach Paleozoic or other fossil coral morphology as a guide to depositional conditions. Normal sedimentologic criteria are likely to be far more reliable, and the presence of coral colonies in a fine-grained or muddy matrix probably signifies accumulation in a quiet water environment, however massive those colonies appear to be.

Two other points need to be made with reference to the Seychelles reefs. First, the comparatively level floor seaward of them is covered with a variety of sediments, ranging from rather poorly sorted muddy gravels in some channels, commonly extensively burrowed by callianassids, to clean-washed sands and coarse gravelly areas. Except on channel floors, corals are not excluded from these environments. Indeed, some of the most prolific growths around the islands occur on comparatively level sandy floors, unrelated to structures which might be termed reefs. Yabe and Sugiyama (1932) described similar coral patches in northern Japanese seas. In such a situation environmental gradients are slight or absent and organisms do not have any obvious zonal distribution. The second observation is that, in addition to these "corals without reefs," there are also present rocky platforms which would be called reefs but which bear no corals! A similar paucity of corals on some reef fronts has been noted in East Africa (J. D. Taylor, personal commun.) and in the Solomon Islands (Stoddart, 1969b).

The foregoing account has tended to emphasize the contribution of corals, as they are among the organisms whose structures are most prominent. It is, however, important to appreciate the very small proportion of the reef which these may represent.

Evaluation of Recent Reefs

From the above data several features can be identified which are common to all recent reefs.

1. They have a topographic expression. In fact, to be "a reef" in the sense of popular usage (irrespective of coral or other connotations), the structure *must* be a ridgelike or banklike form shallow enough to be a hazard to navigation (*cf.* Oxford English Dictionary).

2. This relief is sufficient to cause significant variation in the physicochemical environment around the structure. The rates of change of conditions of lighting, oxygenation, temperature, salinity, turbidity, and similar factors are probably only so rapid in shallow water, bringing the structure close to the air-water interface.

3. Such steep gradients of environmental change inevitably result in a zonation of organisms adapted to specific microenvironments, or limited by the competitive growth of organisms adapted to such environments. Such zonation is primarily a response to tide levels, but the depth to which the reef is affected also depends upon wave power. Thus, structures developed in sheltered environments are less conspicuously zoned.

4. The fourth criterion is less easy to define as it concerns the kind of organisms present. From

the geologic standpoint greatest emphasis must be placed on two particular groups. First are those organisms which act in a constructional, sediment-binding capacity. These, dominated by corals and certain groups of calcareous algae, are responsible for the growth frame, universally recognized as forming the edge zone of existing reefs. The second group is made up of vagrant organisms which secrete more or less solid calcareous skeletons. These can produce only loose sediment, but are of considerable volumetric importance in modern reef environments. The contribution of these elements has been discussed by Chave *et al.* (1972). It is important to note that frame-building organisms not only must be present, they must also be shown to have built a frame. This particular point was contested by Newell (1971) who took a somewhat broader view.

These criteria so far omit any mention of what might lie within the reef. Descriptions at present available in the literature may include reference to the fact that "the reef" is "growing," but apart from comment on control by subsidence and/or sea-level changes they generally go no further. Nor, so far as the neontologist is concerned, is there any need to do so (*cf.* Clausade *et al.,* 1971). It is sufficient, as we have noted, to regard the reef as a passive rocky platform upon which the coral fauna and algal flora represent the equivalent of a climax vegetation. This analogy is also discussed by Newell (1971).

In this view, parallels may be drawn between tropical "coral" reefs and "reefs" in temperate waters of similar profile but bearing instead of corals dense growths of alcyonarians, sponges, and algae. Growth is not necessary to provide the morphology nor, conversely, is the morphology necessary to support the growth; it simply provides a convenient substrate. This may at first seem unconvincing, but in the Seychelles, as noted above, there are both "reefs" without corals and areas of prolific coral growth which are not associated with reefs. Similar coral growth divorced from reef construction was described by Yabe and Sugiyama (1932) from northern Japanese seas.

Reef growth in the West Indian province generally is localized on preexisting topographic highs which project above the general level of sedimentation on the floor of the surrounding shelf (see Logan, 1969; Macintyre, 1972). Macintyre noted that a personal communication from Purdy (in 1971) indicated a thickness of 18 m of Holocene limestone on reefs on the British Honduras shelf, and Logan (1969) believed (without borehole evidence) that relief ascribable to contemporary growth in Yucatan reefs might be of the order of 20-40 m, reached in a period of about 8,000 years. However, these figures may be exceptional. Newell and Rigby (1957) suggested that the Florida reefs are probably a thin skeletal veneer overlying the Key Largo Limestone, and Tracey *et al.* (1948) believed that the reefs of Bikini were growing on an older surface. Even in Yucatan (Logan, 1969) the main reefs are localized on inorganic banks.

Maxwell (1970) described reefs in the Great Barrier Reef complex which owe their morphology to the existence of deltaic salients on which coral communities were later established, and Laborel (1969) has shown that reefs on the Brazilian coast also depend for their form on preexisting bases, at least some of which are sandbars cemented during low sea levels. The importance of other antecedent bases has been apparent in papers by Vaughan (1919) and more recently by Guilcher (1971), Mergner (1971), Stanley and Swift (1968), Montaggioni (1970a, b), and others. Goreau (1961) made the important distinction between coral growth *per se* and the growth of coral reefs, two very different processes which are all too commonly treated as identical.

Estimations of the rates of vertical-reef generation vary and include: Thurber *et al.* (1965, Eniwetok), 36 ft in about 6,000 years; Lalou *et al.* (1966, Mururoa), 23 ft in 2,400 years and 26 ft in 5,000 years; and Purdy (in Macintyre, 1972, British Honduras), 18 m in 3,500 years. The significance of these points will be discussed hereafter.

Fossil "Reefs"

The literature on fossil "reefs" is as large as, if not larger than, that on modern structures. The bibliography of organic reefs, bioherms, and biostromes (Pugh, 1950) was published 20 years ago and would be enlarged considerably if it were revised today. Sanders and Friedman (1967) would be one source of additions to it.

It is not possible to summarize interpretations of fossil reefs by reference to a particular study. This is largely a result of the great diversity of structures involved, and also of a lack of general agreement on what a reef might be.

Dissension and, indeed, confusion have been the hallmarks of studies of fossil reefs almost since their inception. The early controversies of Darwin, Daly, and others were essentially about recent reef structures and later workers have been equally unable to reach a consensus. In 1928 Cumings and Shrock tried to clarify matters by defining two new words (Cumings, 1932)—"bioherm" for "a reef bank or mound . . . of strictly organic origin, embedded in rocks of different lithology," and "biostrome" for "purely bedded structures such as shell beds, crinoid beds, coral beds, *etc.*, consisting of and built mainly by sedentary organisms." Growth frameworks occurred in both groups, but there was no insistence on their presence. Definition was therefore principally on the basis of shape and lithologic contrast. "Bioherm" proved to be a useful term if used in a purely descriptive sense, but its wide application has gathered to it implications which were not originally intended and which have gradually clouded its meaning. Link (1950), for example, regarded it as synonymous with "coral reef" and Taft *et al.* (1968, Fig. 28) appeared to think that it need have no relief at all. Attempts at discussion, as in the Reef Symposium of The American Association of Petroleum Geologists (Wilson, 1950—the opening paper in the symposium volume) have tended to result in a gathering together of divergent opinions rather than any useful agreed definition. The continuing diversification of nomenclature and variation in usage have been summarized by Nelson *et al.* (1962).

As there is no consensus, one way of reviewing supposed paleoreefs is by reference to specific examples. These are all well known, in the sense that their description as reefs is long established, but they vary considerably in the amount and kind of information published about them.

Among the best documented examples are the classic syntheses of the Permian Guadalupe Mountains complex (Adams and Frenzel, 1950; Newell *et al.*, 1953). Basinal deposits, reef limestones, and backreef sediments were recognized, and contrasted in lithology, structures, and faunal characteristics, but it is important to note that no detailed work on the biota had been completed at the time of publication. Unfortunately, this area has suffered also from the inconvenience which seems to have overtaken many fossil reefs, namely, a dolomitization which effectively obliterated structures. Stromatolites and other algal deposits, sponges, and massive bryozoans were described as the chief frame-building organisms.

There is little doubt that this "reef" had real topographic expression and that it controlled the distribution and character of sediments and organisms in the region. What does seem questionable is that this was an organic reef in which bioconstruction determined the feature. Recent papers by Dunham (1969, 1970), Achauer (1969), and Kendall (1969) have shown that certain features, notably so-called "algal" pisolites and the occurrence of frame-building organisms may not be what they appeared to earlier workers. Dunham (1970) stated that the structure was in fact largely the product of inorganic binding, and that organisms, although present, did not provide a rigid framework. In his 1969 paper he also showed that the pisolites, far from being algal, are probably subaerial structures formed in caliche deposits.

Kendall (1969) gave a more detailed facies interpretation along broadly similar lines but with greater emphasis on the supratidal, inorganic origin of many of the sedimentary features. These criticisms and reinterpretations have not gone uncontested (Cys, 1971), but the fact that dispute is possible shows that even the details of evidence, quite apart from the interpretation, are matters of subjective assessment.

The Niagaran "reefs" of the Great Lakes area in the United States present a slightly different problem, although there is also a subjective disagreement on their interpretation. They were described by Cumings and Shrock (1928) and, in greater detail, by Lowenstam (1948, 1950, 1957). The core rocks, where visible and undolomitized, appear for the most part to be fine-grained bioclastic sediment, in which the functions of binding and construction are served by stromatoporoids and *Stromatactis*-like organisms of unknown affinities (Lowenstam, 1950). Descriptions of the structures were hampered by "inadequate exposures and large scale dolomitization" (Lowenstam, 1950, p. 430), but there seems to be a marked faunal difference between "reefs" and "interreef" areas. Relief is varied and Lowenstam recognized three zones: a clastic-free belt in which structures were supposed to rise "less than 10 ft above their surrounding bottoms," an intermediate zone in which they were believed to rise about 40 ft, and a border tract (close to a high-clastic zone) where heights are indicated as of the order of 200 ft. Many of the structures were certainly sites of abundant organic growth and show a faunal progression which was equated with a transition from below to above effective wave base, with consequent increasing turbulence.

Cumings and Shrock (1928) appear to have equated the size of structureless masses of rock visible in present outcrops with relief at the time of deposition. What was their evidence? In spite of the volume of Lowenstam's (1950, 1957) work devoted to reasoning about what a reef might or might not be, relatively few rock descriptions are free of interpretative nomenclature. Lowenstam's own model (1950, 1957) of a shallowing-upward sequence was not met by the material available, "None of the three structures noted above show vertical changes required by this oversimplified pattern of evolution, owing primarily to inadequate vertical, but also unsatisfactory horizontal, exposures, which are limited to parts of the flank deposits" (1950, p. 451). One somewhat odd conclusion reached by Lowenstam on the visible evidence was that wave resistance was implied, *ipso facto,* by all bioclastic debris, and that the presence of this could be used alone as evidence that a reef was present. He used this "fact" (1948) when he suggested that the upper surface of the marine pool "bioherm" was a "reef," even though the rock beneath the bioclast sand was wholly dolomitized.

The particular interest of the preceding interpretation is that Lecompte (1938) already had suggested that this so-called "reef-debris" was not in fact waterworn and, in common with Roll (1934), had decided, after examining the structures, that they were not true reefs. Once again the same evidence is open to different interpretations and the contemporary worker, relying upon fragmentary published descriptions, is unable to be sure exactly what his predecessors saw.

For many examples (but not necessarily for those described above) the interpretation seemingly came first on one set of premises, and the evidence to support it was then recognized in the material available. This was not perforce from any intellectual dishonesty; geologists are trained to make intuitive advances, but some have moved ahead of their facts.

The spur to recognize fossil reefs came initially from their investigation in modern seas and, at the time when their study first began, the simple presence of corals in a limestone sequence was felt to be evidence of "a reef." This seems to have been the case in the Middle Devonian "reefs" in South Devon, described by Worth (1888) as being closely comparable to reefs in the Solomon Islands, a very specific comparison. Unfortunately, once established, such ideas tend to persist, and in subsequent work up to Taylor's (1951) paper these rocks were consistently described as "a reef." The writer (Braithwaite, 1967) demonstrated that, although many corals and stromatoporoids are present, very few are in the position of growth. Most of the sediments are simply bedded bioclastic limestones. It could be argued that these formed part of a reef but that the growth-frame part lay outside existing outcrops, interpreted as talus derived from it. The simpler explanation is that this was a shallow-water, moderate-energy area of high organic productivity, characterized by low rates of other kinds of deposition. In such an environment one might reasonably expect organisms to be broken and disoriented by normal sedimentologic processes. The important difference between these two interpretations is that one requires an act of faith, whereas the other uses only the evidence available in outcrop.

Part of the reason for the persistence of the reef image in this particular area undoubtedly stems from the stratigraphic linkage of the rocks with Middle Devonian carbonates in the Ardennes and Alberta. The Ardennes have become a classic area for "reef" study, largely as a result of the work of Lecompte and his associates (Lecompte, 1938, 1958, 1959, 1970; Tsien, 1971; and many others). Here, ecologic and sedimentologic variations are closely related. The structures themselves, present in the Couvinian, Givetian, and Frasnian, are moundlike forms clearly delimited from the surrounding sediment.

The Frasnian "reefs" are described as being of three kinds: (1) those formed by massive stromatoporoids in the zone of agitation; (2) those formed by corals beneath this zone; and (3) mixed coral and stromatoporoid bodies resulting from subsidence. This interpretation is based on the ecologic assumption that corals could not compete adequately with stromatoporoids in turbulent water and that the latter found it impossible to exist below the zone of agitation.

Emphasis is placed on the importance of *Stromatactis* as a frame-builder, accompanied by laminate corals and stromatoporoids. The writer's studies of the structures indicate that they are largely mud based and cannot, therefore, be expected to have been truly wave-resistant. Lecompte (1959) himself emphasized the lack of frame in some areas, concluding that the structures grew in calm, clay-laden waters. This probably is so, but what connotations are attached to them by labeling them "reefs"? Lees' (1964) interpretation of the Waulsortian banks in Ireland would probably serve as a better model. Textoris and Carozzi (1964) and Lowenstam (1957) also emphasized the frame-building properties of *Stromatactis* but, as Lees (1964) pointed out, the structure need not have been an organism at all. Pray (1958) described *Stromatactis* related to the distribution of bryozoan colonies in Carboniferous bioherms, but Heckel (1972) has shown that many *Stromatactis* were wholly inorganic. It is therefore difficult to ascribe sensibly any frame-building function to them.

Fenton (1931) described Devonian stromatoporoid "reefs" in Michigan. These "reefs" are massively bedded and composed of "lime-sand crowded with broken stromatoporoids," although away from the reefs stromatoporoids increase in size and show neither breakage nor disturbance. Associated with them are thinner muddy limestones and shales containing *Favosites* and brachiopods, and dolomites with *Cylindrophyl-*

lum and *Idiostroma.* No organism is indicated as binding, trapping, or constructing. Recognition of the "reefs" appears to hinge on recognition of potentially (although not actually) constructing organisms, and the presence of a sharp lithologic break rather than an appreciation of ecologic factors involved in distribution. Obviously these facies variations are open to other interpretations.

Layer *et al.* (1949) and Andrichuk (1958) described stromatoporoid-dominated structures in Alberta where three facies zones were recognized in the Leduc "reef," (1) algal or algal-stromatoporoid, (2) stromatoporoid, and (3) nonstromatoporoid. These were compared with recent reef zones by Ladd *et al.* (1950). The algal zone was held to be analogous to the "algal rim," the stromatoporoid zone to the organic lattice, and the nonstromatoporoid zone broadly to backreef environments. Much of the so-called framework appears to have been loose stromatoporoids in a bioclastic matrix. Reconstructed as an atoll-like rim for an area in which there is insufficient data, the individual zones can be more than 1 mi in width with only poorly defined boundaries. Although the zones were very probably shallow-water environments, gradients of change were low and it is possible to explain the distribution without involving "a reef."

Other Paleozoic "reefs," which perhaps are best known to British geologists, are the "reef knolls" of the Carboniferous of Yorkshire. One of the problems in interpreting them has been a lack of adequate exposure of critical areas, and this has resulted in a great diversity of opinions as to origin. Tectonic and various kinds of depositional mechanisms have been suggested, but with one exception, authors have been content with the single word "reef" as a description. Much of the history of observation of these structures is reviewed by Black (1954) and Miller and Grayson (1972)

Much less work has been published on reefs in Mesozoic and Tertiary rocks. Henson (1950) summarized the features of some Middle Eastern structures and Forman and Schlanger (1957) extended some of the work begun by Vaughan (1919) in the southeastern United States. Griffith *et al.* (1969) made an interesting comparison of Cretaceous sediments in Mexico with recent environments in Florida, and Arkell (1928, 1935) compared Jurassic environments in England with those in the Red Sea. Arkell (1928) in particular gave little information on the structure of his "reefs," although he made very detailed descriptions of his interpretations of reef development (1935).

One of the more disturbing aspects of the reef problem is illustrated by Wilson's (1949) account of a small "reef" in the Jurassic of the Hackness hills in Yorkshire. The maximum development of the unit was about 11 ft. Unfortunately, most of the outcrops are now obscured by infilling and are not open to reexamination. The only records are those in which rocks are described as "reef rock" or "reef-talus" and there is no way of knowing what these terms meant. *In situ* corals still are visible in quarries, but there is no way of knowing whether these were ever more than a few colonies growing on a shallow floor. Wilson (1949) did not describe the relation of the colonies to each other or indicate the basis for his belief in a "reef."

Relevance of Structure and History of Recent Reefs

Internal structure is among the most difficult aspects of the study of recent reefs. The literature is full of statements to the effect that reefs are "growing" in a certain direction. So they may be, in the sense that that is where growth of corals appears to be taking them, but was that growth solely responsible for the form now visible?

Every student is familiar with Darwin's (1890) model of a subsiding island. The corals growing around it, occupying the favorable environments of maximum water circulation, form a rim which grows upward to sea level before riding outward over an apron of its own debris. This model, in spite of its qualification by Daly (1915), Ladd and Hoffmeister (1936), and many others (see Davis, 1928; Gardiner, 1931) has remained viable as a textbook illustration. It is a simple paradigm which can be modified to explain the effects of different rates of subsidence or senses of sea-level change (as in the models of Davis, 1928; Cloud, 1952). One important factor is commonly neglected, although Davis seems to have understood its significance. The history of recent reefs, and probably that of many of the structures with which they are being compared, is not marked either by a simple rise or fall in sea level (be it epeirogenic or eustatic), but by oscillation.

The effects of oscillation can be examined only with difficulty. Some available reports on boreholes through carbonate sequences below reefs are the following:

Bahamas: Field and Hess (1933)
Bikini: Emery *et al.* (1954)
Eniwetok: Ladd and Schlanger (1960); Schlanger (1963)
Funafuti: Cullis (1904); Judd (1904)
Great Barrier Reef: Richards and Hills (1942)
Hawaii: Lum and Stearns (1970)
Kita-Daito-Jima: Yasushi (1939, in Japanese); some results in Ladd *et al.* (1970)
Midway: Ladd *et al.* (1967, 1970)

Mururoa: Chauveau *et al.* (1967)

It seems inevitable that many of these investigators were influenced by the fact that they knew in advance that they were boring through a coral reef. Their samples thus were interpreted in relation to the presence somewhere of such a structure. In some cases this was justified by the recognition of diagnostic corals, Foraminifera, or other organisms, but much of the Bikini core, for example (Emery *et al.*, 1954), consists of bioclastic sediment, which is unconsolidated and indicates only that the particular organisms were growing in the area (see also Ladd *et al.*, 1970). The presence of erosion surfaces in the Eniwetok cores (Schlanger, 1963) and of lignitic layers on Midway (Ladd *et al.*, 1970) is significant in indicating that deposition was not continuous. These conclusions may be augmented by studies of surface outcrops over a more limited time range, such as the following.

Barbados: Mesolella *et al.* (1970)
Bermuda: Land *et al.* (1967)
Florida: Stanley (1966)
Hawaii: Lum and Stearns (1970)
Madagascar: Battistini (1969)
Sudan: Sestini (1965)

Land and Goreau (1970) and Ginsberg and Schroeder (1969) tried to investigate reef structures by underwater excavation. Ginsberg and Schroeder used explosives to produce sections 8 m high and were unable to find either an antecedent base or bedding within the reef. More work of this kind is needed, using techniques to penetrate deep within the sediment body.

The late Pleistocene succession on Aldabra (Indian Ocean) serves as an example (Braithwaite *et al.*, ?1973). The wholly carbonate sequence results from at least 12 separate events and records the development of the island during a substantial period. Events include high sea-level stands, indicated by marine sediments; low sea-level stands, represented by terrestrial deposits; and marine and terrestrial erosion surfaces. The terrestrial faunas of mollusks, tortoises, crocodiles, birds, and lizards record a pattern of successive colonizations of the newly emergent land surfaces.

Two critical points can be made about this sequence. The first is that here, as in many other areas, the volume of rock in outcrop occupied by coral growth frame is small. Growth frame is not, of course, dominant in recent reefs, but in this case (as in the Pleistocene on the East African mainland) it has been further reduced by erosion. This probably will be so in any situation where growth is followed by emergence. A second point is that emergence has taken place on several occasions in the observed history of the islands. During these erosional intervals morphology is redefined and later submergence results in the colonization of favorable areas by frame-building corals. Therefore reef growth in these cases, far from being an active organic process, might be viewed as a passive accretion of successive units of sediment, albeit strongly influenced by the growth of organisms. This is not a new idea; some elements of it clearly belong to Daly (1915), to Hoffmeister and Ladd (1935), MacNeil (1954a, b), Stoddart (1969b), and others. MacNeil (1954a, b) put forward the idea that much of the morphology of modern reefs is explicable in terms of growth of corals on a preexisting base, the shape of which was largely responsible for the form which the reef later took. His illustrations were related particularly to karstic landforms produced on the emergent limestone surface during lowered sea level. Dodd and Siemers (1971), Goreau and Burke (1966), Guilcher (1971), Laborel (1969), Montaggioni (1970a, b), Stanley and Swift (1968), and Weydert (1969) have similarly shown (in different areas) how form and even distribution of the biota may be controlled by Pleistocene and more recent erosion features.

The Seychelles reefs lend support to this idea in so far as recent sediment accumulations appear to be in the process of building up to, but have not yet reached, sea level. As the writer (Braithwaite, 1971) has shown, reef channels are not now primarily axes of sediment movement, neither do they seem to be barren residuals of an advancing reef front. Three-dimensional studies in boreholes and large-scale excavations show that the reef edges have originated largely on their present sites and are not the products of seaward migration of corals which have reached their limiting sea level. The fact that the sediments are unconsolidated presents no problem, for so are most of the carbonate sediments on Aldabra and the East African coast. Excavations show one additional feature of importance, the sediments, at least in the backreef area, are bedded. Lithologic continuity is, however, limited and individual units can be traced only over small areas.

In summary, present reefs commonly are bedded structures. This is borne out by excavations in recent sediments in the Seychelles, observation of the Pleistocene of Aldabra and of the East African coast, and examination of the boreholes listed earlier. Coral frames commonly represent only a small part of the volume. Recognition of "the reef" seems to require information on the morphology and nature of the foundation on which the growth frame rests as well as the plant and animal communities forming it. The last two points seem to be of particular significance to borehole studies of ancient rock sequences.

Criteria for Reef Recognition

After the preceding summary of the characteristics by which reefs are recognized in modern environments, and of some of the deficiencies of "reefs" in ancient stratigraphic sequences, it is desirable to examine these criteria to assess the limits of their applicability.

Relief—To determine relief requires the identification of some kind of intraformational surface having the necessary irregularity. There may be a lithologic discontinuity, but this alone is not sufficient evidence, and such a surface could be present with little or no lithologic differentiation. Evidence of a well-defined growth frame; crusting by calcareous algae, sponges, or similar organisms; or boring by sponges, bivalves, or other organisms may indicate that the surface was hard and would be of obvious value, but what must be shown is that the structure had elevation at the time when it was a site of active coral growth. The scale of relief is important, because it determines the environmental gradients controlling ecology. The limiting size is admittedly subjective, depending partly on continuity. A mass of corals rising a few meters above the sea floor might be called a reef if it reached the surface and extended for 100 m or more, but is (presumably) not a reef if only one or two colonies are involved. "Reefs" such as that noted by Ager (1963) built by annelids to a height of 10 cm, although separated from "real reefs" by their inclusion in inverted commas, are clearly of a quite different order of magnitude. Similarly, Gram's (1968) and Kirtley and Tanner's (1968) descriptions of sabellarid "reefs" seem to me unacceptable, because, although they apparently have considerable lateral continuity, their average height is only a little over 100 cm. However, reefs are not the only structures which, even in shallow water, rise significantly above the sea floor. Mud banks in Florida Bay (Ginsberg and Lowenstam, 1958), in Shark Bay, West Australia (Davies, 1970), and in the Mediterranean (Molinier and Picard, 1952) are all the products of organic binding, but by marine phanerogams (*Thalassia* and *Posidonia*) rather than by any organism with a rigid mineral skeleton. The Florida examples have a relief of 2-2.5 m, and those in the Mediterranean may be up to 6 m, having accumulated within about 1,200 years. In Shark Bay the grasses are associated with nodose and crustose calcareous algae, but these play a minor role in sediment binding. On the Alacran reef (Logan, 1969), banks within the lagoon have a relief of from a few feet to perhaps 50 ft and appear to have been formed primarily by the binding activities of marine grasses. The biota includes calcareous algae and a variety of corals, but there is little or no tendency for them to form rigid or semirigid frames.

Riding and Toomey (1972) discussed the role of certain algae in the construction of some Ordovician mud mounds and showed that they also functioned without a rigid frame. Other banks such as those on which the Yucatan reefs are founded (Logan, 1969) may be inorganic in origin. They might be recognized in a fossil sequence as being sharply differentiated from surrounding sediments. They may be diagenetically altered (dolomitized) and obviously may have formed topographic highs, perhaps even surrounded by "reef detritus," but these facts alone do not constitute a reef.

Environmental variation—The change in environments around a reef is clearly significant. As indicated earlier in the discussion, this condition can be met adequately only in shallow waters. Probably all coasts are sedimentologically and faunally zoned, as is any place where physical processes combine to produce an environmental gradient (*e.g.*, see Walker, 1972), but the rapid changes around reefs can be found only offshore on shelflike features reaching to near sea level. The fact that the conditions are universally determined by the morphology is shown by comparing Maxwell's (1970) delta-based reefs with structures which are presumed to lie on a more conventional coral-limestone base.

Zonation—The high rates of change of environments and the juxtaposition of contrasted habitats are responsible for the commonly abrupt variations in reef biota which result in a prominent zonation (see Stoddart, 1969a). Without zonation, which is commonly rather precise in its boundaries, a reef probably is not present. The absence of such zonation on deep-water coral banks (Teichert, 1958; Stetson *et al.*, 1962; Maksimova, 1972) has been noted, and fossil structures lacking zonation probably were below the limits of surface interference. Maksimova (1972) also made the significant point that deep-water structures lack the association with bioclastic sediments formed by wave action.

Nature of organisms—If one accepts a "reef" defined on the premises already outlined, there is no good reason for limiting the ecosystem to the groups of organisms which happen to be dominant at the present time. However, they should be analogous in function. That is to say, they should include sediment-supplying organisms, but be dominated by those which, from branching or encrusting habit, are potentially capable of constructing a rigid frame.

The most obvious fossil group to which this provision might apply is the stromatoporoids, which produce a skeleton of general form similar to that of some modern corals. Massive colonies, digitate growths, and encrusting sheets are known among fossils and might be compared morphologically with colonies of, for example, species within the genus *Porites*. They also reach an order of size comparable with that of some corals, and are controlled environmentally.

Until recently, stromatoporoids were considered to be somewhat problematic and could be assigned to almost any environment. The work of Hartman and Goreau (1970) leaves less freedom. Stromatoporoids appear to have been sponges, and modern genera inhabit clear waters, although they generally avoid strong light by living in caverns or at depth (70 m in Jamaica). In deep water they compete with a fauna which includes other coralline sponges and hermatypic corals. A similar association has been described on Madagascar (Vacelet and Vasseur, 1971). The apparent correlation of some fossil stromatoporoids with muddy sediments may be a reflection of this photophobic habit, suspended sediment reducing light intensity and permitting growth in shallower water, but their rounded shape certainly does not appear to be related to any wave resistance.

Less robust organisms cannot be described so readily as "potentially frame-building." The most widely quoted group which does not meet this requirement is the filamentous algae. "Reefs," supposedly built by these organisms forming stromatolites, have been reported from several areas (Kruger, 1969; Smith and Francis, 1967). Smith and Francis (1967) described a Permian "reef" in the Durham and Hartlepool area which was in parts "almost wholly of algal (stromatolitic) origin," and the reef face was reported to be ultimately over 200 ft high with a slope of between 30 and 85°. In the light of the now generally accepted accounts of algal stromatolites of Black (1933) and Logan *et al.* (1964), there seem good grounds for suggesting that the sediment relations are not interpretable in this way. What kind of unlithified sediment, even if bound by filamentous algae, is capable of resting on such a slope, close to the surface, even assuming that the algae were able to build it selectively? It also seems a coincidence that sediment being generated in, or carried to, the area should conveniently stop at the edge of the slope, and strange, in view of the organisms present, that associated evaporites should be present in the basin rather than on the adjacent shelf.

The *Thalassia*-bound banks of Molinier and Picard (1952) and Davies (1970) have already been mentioned. This kind of structure shows a fundamental difference in the mechanics of growth from that of reefs and relies on a process of passive accretion of loose sediment rather than active construction of a frame, although the organisms concerned are clearly vital in bringing about that accretion.

Banks or ridges formed solely by the growth of successive generations of oysters have been described by Norris (1953) and (in summary) by Stenzel (1971, p. N1041-N1048). Although these banks are constructional in the sense that they are built by sessile skeleton-forming organisms, their monospecific character and lack of zonation differ from the reef model and there is little point in concealing this difference by calling them reefs.

The reef biota has undergone many changes in the history of these structures. Obviously the organisms available to form reefs have been changing, but so have the vagrant genera associated with them. Notable among these latter changes is the progressive exploitation of the infaunal habitats by bivalves (Stanley, 1968) and the replacement of brachiopods by epifaunal members of the same group. Other evolutionary effects such as the extinction of the trilobites and rise of the decapod crustaceans have influenced both the sediments and the structures within them. Probably in many physically similar environments since the Paleozoic, mechanically induced structures have been largely replaced by those of biogenic origin. Boring organisms also have steadily increased in number and importance.

Internal structure.—The most important conclusion drawn from the discussion of recent structures is that they commonly are bedded (albeit crudely in some cases) and have not been raised as a single unit. This would not be the case if sea level were changing in a constant direction and at a constant rate, so that depth and the physical characters of the environment remained the same.

It is arguable that sea level in the past may have shown this kind of consistency of direction, but we are making an assumption. Present processes seem *potentially* capable of producing present morphology, but it is too easy to move from that position to the belief that they have in fact done so. The internal structure of a reef indicates the constancy (or lack of it) of a particular set of environmental conditions, and the recent history of sea-level changes has been marked by oscillation.

Sedimentologic Models

The groups of diagrams presented are intended

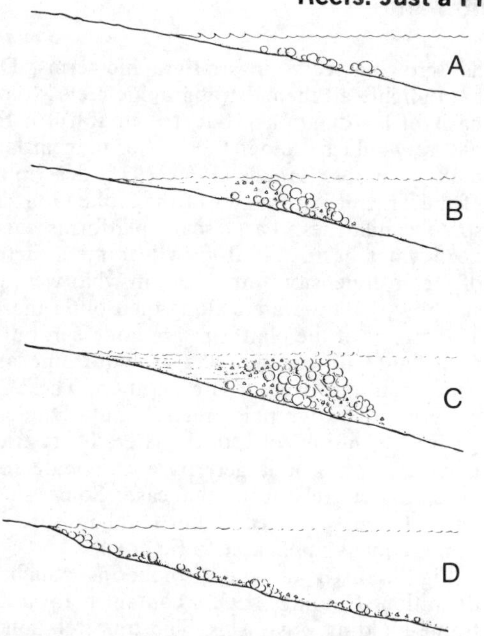

FIG. 1—Sections illustrate classical concept of reef growth: **A,** corals are established on virgin surface; **B,** they grow in relation to stable sea level; or, **C,** rise slowly with accretion taking place to sea level and in seaward direction; **D,** sea level has been assumed to rise so fast that coral growth was unable to keep pace; corals at lower levels die and are overwhelmed by sediment. See Figure 2 for legend.

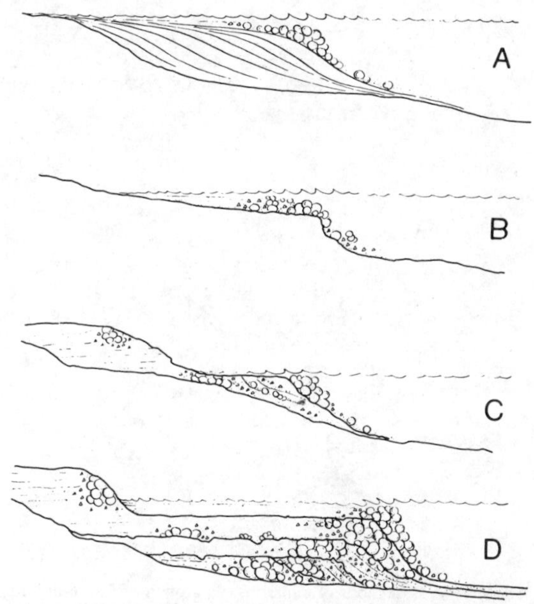

FIG. 3—**A,** section of reef growing on antecedent base formed from old delta; **B-C,** sections of reefs on antecedent bases formed from erosional benches of differing origin (see text); **D,** section of reef in which base has had complex history of submergence and emergence. See Figure 2 for legend.

to portray structures which have been described as reefs. These should make clear the basis of the writer's thesis and avoid any possible ambiguity which might arise from the necessarily brief descriptions given previously.

Figures 1A-C and 3A-C are of structures which would be recognized as "reefs" in modern environments, and illustrate how essentially similar morphologies may result from very different patterns of depositional history. Figure 1A-C may be entirely theoretical. It defines the classical concept of coral growth established on a virgin surface (A), growing up in relation to a stable sea level (B), or one which is rising slowly with accretion taking place in a seaward direction (C). At stage A coral growth is of insufficient relief to be called "a reef." Figure 1D presents the position in which sea level is assumed to have risen so fast that coral growth was unable to keep pace, corals at lower levels dying off and being overwhelmed by sediment.

Figure 3A-C represents growth on some antecedent base. Figure 3A is a situation similar to that described by Maxwell (1970), where growth is superimposed on a tidal-channel delta. Figure 3B is broadly the case in East Africa where corals have colonized the margin of what is largely a solution and abrasion bench in Pleistocene limestones. Figure 3C, a theoretical model (which

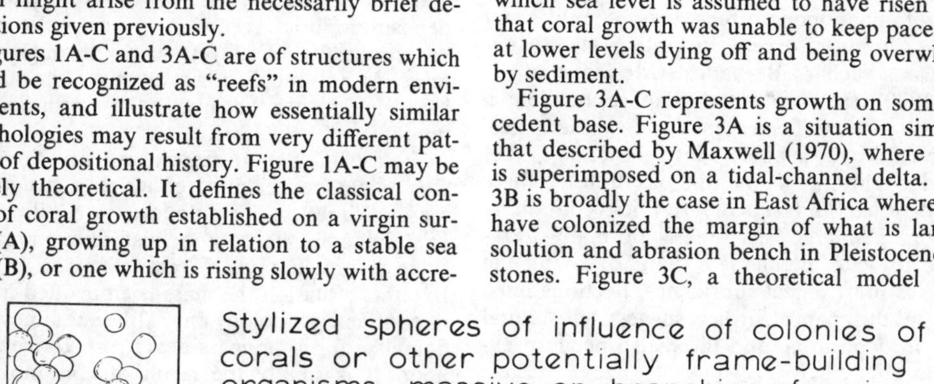

FIG. 2—Legend for all figures. *Correction:* calcrudite should be calcirudite.

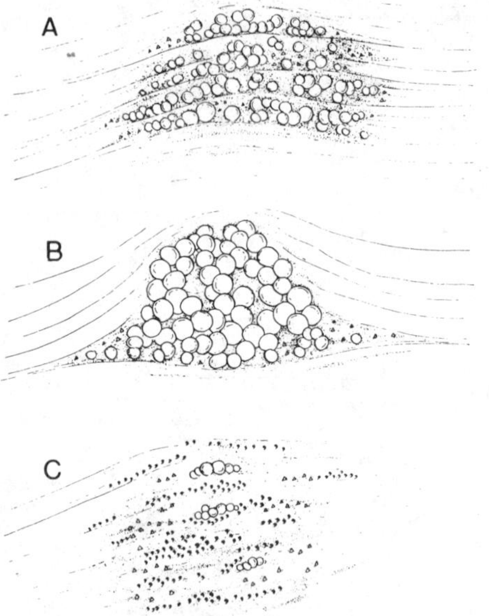

FIG. 4—**A-B**, sections of bioherms: **A**, growth of frame-building organisms occupied same site over extended period, but at no time formed structure with significant relief; **B**, growth frame was built up above surrounding sea floor; **C**, cross section of laterally limited mass of limestone formed by growth of particular organisms in one area during extended period. With little modification, model indicates relations where large organoclasts are not involved or where core is dolomitized. See Figure 2 for legend.

may be true for East Africa), illustrates the continuation of such a process where sea level falls again.

In fact, many modern reefs may be unlike any of these, because, except in certain clearly defined cases such as Barbados, (Mesolella *et al.*, 1969, 1970), they are believed to result from a repetitive pattern of emergence and submergence. Even on Barbados, where Mesolella *et al.* (1970) showed a pattern of "continuous uplift," there are individual units 60-100 ft thick containing *Acropora palmata*. This species normally is restricted to water depths less than 25 ft, and the units thus must reflect subsidence. Sections illustrated in this paper further suggest substantial breaks in deposition, and these also need an explanation.

In general terms, a history of repeated immersion and emersion might be expected to produce a structure resembling Figure 3D. The contemporary stage of this is "a reef" (in so far as it is described as such) and so, presumably, were some of the earlier stages, but it cannot be described in quite the same terms of "reef growth" as Figure 1C which is an "ecological reef" in the sense of Dunham (1970).

Also worth considering here is whether features shown in Figures 3A-C would be regarded as "growing reefs" in stratigraphic terms. Dunham might call them "stratigraphic reefs" (on the basis of his diagrams), but it is doubtful if their nature would be understood other than in large-scale outcrops. Zankl's (1969, 1971) descriptions of the Triassic carbonates of the Hohe Goll show some similarities. Large-scale platforms face a deep-water basin. Coral growth-frame structures of meter dimensions are found in what were presumably shallow waters along their outer margin. The origins of the platforms are not clear, but the important point is that the growth-frame areas were small and were a late addition. These may well be "stratigraphic reefs," but Dunham's (1970) definition of "thick laterally restricted masses of pure or largely pure carbonate rock" does not entirely meet the case. Some similar term obviously is needed, but Dunham's description is equally applicable to Figure 4.

Figure 4A-B illustrates bioherms which, by definition (Cumings, 1932), contain many *in situ* frame-building organisms. The time relation between the structures and their surrounding sediments is critical, as similar facies relations might result from the accretion of a mound, later surrounded by sediment (B), or simply by the persistence in one area of a particular kind of deposition (A). The latter situation is clearly that envisaged in Scoffin's (1971) paper, which can be recommended for the descriptions of relations between organic growth and surrounding sediments. Once established, growth kept pace with deposition, brief recessions being indicated by the continuity of clayey seams throughout the mass. At no time during growth did the structure have significant relief above the surrounding sea floor.

Figure 4C demonstrates two situations. The core might be imagined either as a structureless mass of limestone or dolomite, or as (illustrated) containing fossils of nonconstructing organisms such as brachiopods, foraminifers, or trilobites. In either situation the mass is embedded in rocks of contrasted lithology and, although it now has a domelike form the mass never rose above the sea floor. It is simply the result of continuity of a particular kind of deposition in a limited area.

Hadding (1950), describing Silurian "reefs" in Gotland, noted that they contained no large resistant calcareous bodies such as stromatoporoids or corals. They were deposited in quiet waters and probably could not maintain their growth without the support of new sediments deposited on their sides.

Obviously the interpretation of many of these structures is going to be difficult, and must be made with particular care where outcrops are limited or where data are based on boreholes, but it contributes little to our understanding of this multiplicity of structures to obscure them under the single title "reef." The economic advantages of differentiation are also clear. The character and distribution of porosity are important in controlling the occurrence of petroleum or other resources within these structures (*e.g.,* Stanton, 1967; Gerdemann and Myers, 1972). A proper awareness of the factors determining porosity distribution is therefore necessary both to the discovery and exploitation of such resources.

Conclusions

The criteria by which fossil reefs can be recognized are four.

1. The structure can be shown to have risen appreciably above the sea floor at the time of maximum organic growth.
2. The structure was wave resistant, lying within the zone of wave action.
3. The organisms present included frame-building forms which can be shown to have functioned as frame-builders.
4. Both the frame-builders and associated vagrant organisms show a zonation consequent upon their near-surface position.

This definition clearly leaves the way open to recognizing more than one kind of reef. First are those structures which have gained their elevation and wave resistance solely by the growth of frame-building organisms; these are perhaps rare. Second are those structures in which part of the relief is provided by an antecedent platform. Both are present in modern seas, and might be termed respectively biogenic and morphologic reefs. Such terms however, should be used only in summary and not in primary descriptions.

Only when the conditions attached to all these points are met can the sediment body be described as a reef analogous to modern structures. Where these conditions are not present, some other kind of structure is involved. The latter undoubtedly exist, but are commonly disguised, because of their appearance under the misleading title of "reef," particularly where they lack any detailed description.

Unless geologists provide adequate descriptions, *followed* by interpretations, we can look forward only to further years of dispute. The recent "reef" is the only model that can be used for structures which, by the use of this word, are being gauged against it.

References Cited

Achauer, C. W., 1969, Origin of Capitan Formation, Guadalupe Mountains, New Mexico and Texas: Am. Assoc. Petroleum Geologists Bull., v. 53, no. 11, p. 2314-2323.

Adams, J. E., and H. N. Frenzel, 1950, Capitan barrier reef, Texas and New Mexico: Jour. Geology, v. 58, no. 4, p.289-312.

Ager, D. V., 1963, Principles of paleoecology: New York, McGraw-Hill, 371 p.

Allen, J. R. L., 1971, Instantaneous sediment deposition rates deduced from climbing-ripple cross-lamination: Geol. Soc. London Quart. Jour., v. 127, no. 6, p. 553-561.

Andrichuk, J. M., 1958, Stratigraphy and facies analysis of Upper Devonian reefs in Leduc, Stettler, and Redwater areas, Alberta: Am. Assoc. Petroleum Geologists Bull., v. 42, no. 1, p. 1-93.

Arkell, W. J., 1928, Aspects of the ecology of certain fossil coral reefs: Jour. Ecology, v. 16, no. 1, p. 134-149.

——— 1935, On the nature, origin, and climatic significance of the coral reefs in the vicinity of Oxford: Geol. Soc. London Quart. Jour., v. 91, pt. 1, p. 77-110.

Battistini, R., 1969, Les modifications du climat a Madagascar au Plio-Quaternaire: Acad. Malgache Mem., no. 43, p. 103-114.

Black, M., 1933, The algal sediments of Andros Island, Bahamas: Royal Soc. London Philos. Trans., ser. B., v. 222, pt. 6, p. 165-192.

Black, W. W., 1954, Diagnostic characters of the Lower Carboniferous knoll-reefs in the north of England: Leeds Geol. Assoc. Trans., v. 6, pt. 5, p. 262-297.

Braithwaite, C. J. R., 1967, Carbonate environments in the Middle Devonian of south Devon, England: Sed. Geology, v. 1, no. 4, p. 283-320.

——— 1971, Seychelles reefs; structure and development, *in* Regional variation in Indian Ocean coral reefs: Zool. Soc. London Symp. 28, p. 39-63.

——— and M. R. Talbot, 1972, Crustacean burrows in the Seychelles, Indian Ocean: Palaeogeography, Palaeoclimatology, Palaeoecology, v. 11, no. 4, p. 265-285.

——— J. D. Taylor, and W. J. Kennedy, ?1973, The evolution of an atoll: the depositional and erosion history of Aldabra: Royal Soc. London Philos. Trans. (in press).

Chauveau, J. C., G. Deneufbourg, and J. A. Sarcia, 1967, Observations sur l'infrastructure de l'atoll de Mururoa (archipel des Touamotou, Pacifique sud): Acad. Sci. Comptes Rendus, ser. D, v. 265, p. 1113-1116.

Chave, K. E., S. V. Smith, and K. J. Roy, 1972, Carbonate production by coral reefs: Marine Geology, v. 12, no. 2, p. 123-140.

Clausade, M., N. Gravier, J. Picard, M. Pichon, M-L. Roman, B. Thomassin, P. Vasseur, M. Vivien, and P. Weydert, 1971, Coral reef morphology in the vicinity of Tulear (Madagascar): contribution to a coral reef terminology: Station Marine d'Endoume, Tethys Supp. 2, p. 1-76.

Cloud, P. E., 1952, Facies relationships of organic reefs: Am. Assoc. Petroleum Geologists Bull., v. 36, no. 11, p. 2125-2149.

——— 1959, Geology of Saipan Mariana Islands; pt. 4, Submarine topography and shoal-water ecology: U.S. Geol. Survey Prof. Paper 280-K, p. 361-445.

Cullis, C. G., 1904, The mineralogical changes observed in the cores of the Funafuti borings, *in* The atoll of Funafuti: Royal Soc. London Rept. Coral Reef Comm, sec. 14, p. 392-420.

Cumings, E. R., 1932, Reefs or bioherms?: Geol. Soc. America Bull., v. 43, no. 1, p. 331-352.

——— and R. R. Shrock, 1928, Niagaran reefs of Indiana and adjacent states and their stratigraphic relations: Geol. Soc. America Bull., v. 39, no. 2, p. 579-620.

Cys, J. M., 1971, Origin of Capitan Formation, Guadalupe Mountains, New Mexico and Texas: discussion: Am. Assoc. Petroleum Geologists Bull., v. 55, no. 2, p. 310-312.

Daly, R. A., 1915, The glacial-control theory of coral reefs: Am. Acad. Arts Sci. Proc., v. 51, p. 157-251.

——— 1948, Coral reefs, a review: Am. Jour. Sci., v. 246, no. 4, p. 193-207.

Darwin, C., 1890, On the structure and distribution of coral reefs: London, Ward, Lock and Co., 154 p.
Davies, G. R., 1970, Carbonate bank sedimentation, eastern Shark Bay, Western Australia, *in* Carbonate sedimentation and environments, Shark Bay, Western Australia: Am. Assoc. Petroleum Geologists Mem. 13, p. 85-168.
Davis, W. M., 1928, The coral-reef problem: Am. Geog. Soc. Spec. Pub., no. 9, 596 p.
Dodd, J. R., and C. T. Siemers, 1971, Effect of late Pleistocene karst topography on Holocene sedimentation and biota, lower Florida Keys: Geol. Soc. America Bull., v. 82, no. 1, p. 211-217.
Doty, M. S., 1957, Rocky intertidal surfaces, *in* Treatise on marine ecology and paleoecology: Geol. Soc. America Mem. 67, v. 1, p. 535-585.
——— and J. P. E. Morrison, 1954, Inter-relationships of the organisms on Raroia aside from man: Atoll Research Bull., v. 35, p. 1-61.
Dunham, R. J., 1962, Classification of carbonate rocks according to depositional texture, *in* Classification of carbonate rocks: Am. Assoc. Petroleum Geologists Mem. 1, p. 108-121.
——— 1969, Vadose pisolite in the Capitan reef (Permian), New Mexico and Texas, *in* Depositional environment of carbonate rocks: Soc. Econ. Paleontologists and Mineralogists Spec. Pub. 14, p. 182-191.
——— 1970, Stratigraphic reefs versus ecologic reefs: Am. Assoc. Petroleum Geologists Bull., v. 54, no. 10, p. 1931-1932.
Edmondson, C. H., 1928, The ecology of an Hawaiian coral reef: Bernice P. Bishop Mus. Bull., v. 45, p. 1-64.
Emery, K. O., J. I. Tracey, Jr., and H. S. Ladd, 1954, Geology of Bikini and nearby atolls: U.S. Geol. Survey Prof. Paper 260-A, p. 1-265.
Fairbridge, R. W., 1950, Recent and Pleistocene coral reefs of Australia: Jour. Geology, v. 58, no. 4, p. 330-401.
Fenton, M. A., 1931, A Devonian stromatoporoid reef: Am. Midland Naturalist, v. 12, no. 7, p. 195-202.
Field, R. M., and H. H. Hess, 1933, A bore hole in the Bahamas: 14th Ann. Mtg. Am. Geophys. Union Trans., p. 234-245.
Finck, A. E., 1904, Biology of the reef-forming organisms of Funafuti Atoll, *in* The atoll of Funafuti: Royal Soc. London Coral Reef Comm. Rept., sec. 6, p. 125-150.
Folk, R. L., 1962, Spectral subdivision of limestone types, *in* Classification of carbonate rocks: Am. Assoc. Petroleum Geologists Mem. 1, p. 62-84.
Forman, M. J., and S. O. Schlanger, 1957, Tertiary reef and associated limestone facies from Louisiana and Guam: Jour. Geology, v. 65, no. 6, p. 611-627.
Gardiner, J. S., 1931, Coral reefs and atolls: London, Macmillan, 181 p.
Garrett, P., D. L. Smith, A. O. Wilson, and D. Patriquin, 1971, Physiography, ecology, and sediments of two Bermuda patch reefs: Jour. Geology, v. 79, no. 6, p. 647-668.
Gerdemann, P. E., and H. E. Myers, 1972, Relationships of carbonate facies patterns to ore distribution and to ore genesis in the southeast Missouri lead district: Econ. Geology, v. 67, no. 4, p. 426-433.
Ginsburg, R. N., and H. A. Lowenstam, 1958, The influence of marine bottom communities on the depositional environment of sediments: Jour. Geology, v. 66, no. 3, p. 310-318.
——— and J. H. Schroeder, 1969, Introduction to the growth and diagenesis of Bermuda reefs: Bermuda Conf. on Carbonates Preprint, 13 p.
Goreau, T. F., 1959, The ecology of Jamaican coral reefs, 1, Species composition and zonation: Ecology, v. 40, no. 1, p. 67-90.
——— 1961, The structure of the Jamaican reef communities: geological aspects: New York Zool. Soc. Coral Reef Project, U.S. Navy Office of Naval Research, Biol. Branch, Final Prog. Rept., Contract (G)0003-60, 14 p.
——— and K. Burke, 1966, Pleistocene and Holocene geology of the island shelf near Kingston, Jamaica: Marine Geology, v. 4, no. 3, p. 207-225.
Gram, R., 1968, A Florida *Sabellariidae* reef and its effect on sediment distribution: Jour. Sed. Petrology, v. 38, no. 3, p. 863-868.
Griffith, L. S., M. G. Pitcher, and G. W. Rice, 1969, Quantitative environmental analysis of a Lower Cretaceous reef complex, *in* Depositional environments in carbonate rocks: Soc. Econ. Paleontologists and Mineralogists Spec. Pub. 14, p. 120-138.
Guilcher, A., 1971, Mayotte barrier reef and lagoon, Comoro Islands, as compared with other barrier reefs, atolls, and lagoons in the world, *in* Regional variation in Indian Ocean coral reefs: Zool. Soc. London Symp. 28, p. 65-86.
Hadding, A., 1950, Silurian reefs of Gotland: Jour. Geology, v. 58, no. 4, p. 402-409.
Hartman, W. D., and T. F. Goreau, 1970, Jamaican coralline sponges: their morphology, ecology, and fossil relatives: Zool. Soc. London Symp. 25, p. 205-243.
Heckel, P. H., 1972, Possible inorganic origin for *Stromatactis* in calcilutite mounds in the Tully Limestone, Devonian of New York: Jour. Sed. Petrology, v. 42, no. 1, p. 7-18.
Henson, F. R. S., 1950, Cretaceous and Tertiary reef formations and associated sediments in Middle East: Am. Assoc. Petroleum Geologists Bull., v. 34, no. 2, p. 215-238.
Hoffmeister, J. E., and H. S. Ladd, 1935, The foundations of atolls: a discussion: Jour. Geology, v. 43, no. 6, p. 653-665.
Jones, J. A., 1963, Ecological studies of the southeastern Florida patch reefs; 1, Diurnal and seasonal changes in the environment: Bull. Marine Sci., v. 13, p. 282-307.
Jopling, A. V., and R. G. Walker, 1968, Morphology and origin of ripple-drift cross-lamination with examples from the Pleistocene of Massachusetts: Jour. Sed. Petrology, v. 38, no. 4, p. 971-984.
Judd, J. W., 1904, General report on the materials sent from Funafuti and the methods of dealing with them, *in* The atoll of Funafuti: Royal Soc. London Coral Reef Comm. Rept., sec. 10, p. 167-185.
Kendall, C. G. St. C., 1969, An environmental re-interpretation of the Permian evaporite/carbonate shelf sediments of the Guadalupe Mountains: Geol. Soc. America Bull., v. 80, no. 12, p. 2503-2525.
Kirtley, D. W., and W. F. Tanner, 1968, Sabellariid worms: builders of a major reef type: Jour. Sed. Petrology, v. 38, no. 1, p. 73-78.
Kruger, L., 1969, Stromatolites and oncolites in the Otavi series, South-West Africa: Jour. Sed. Petrology, v. 39, no. 3, p. 1046-1056.
Laborel, J., 1969, Les peuplements de madreporaires des cotes tropicales de Bresil: Annales Univ. Abidjan, ser. E, v. 2, no. 3, p. 1-260.
Ladd, H. S., and J. E. Hoffmeister, 1936, A criticism of the glacial-control theory: Jour. Geology, v. 44, no. 1, p. 74-92.
——— and——— 1945, Geology of Lau, Fiji: Bernice P. Bishop Mus. Bull., v. 181, p. 1-309.
——— and S. O. Schlanger, 1960, Drilling operations on Eniwetok Atoll: U.S. Geol. Survey Prof. Paper 260-Y, p. 863-905.
——— and J. I. Tracey, Jr., 1949, The problem of coral reefs: Sci. Monthly, v. 69, no. 5, p. 297-305.
——— ——— and M. G. Gross, 1967, Drilling on Midway Atoll, Hawaii: Science, v. 156, p. 1088-1094.
——— ——— and——— 1970, Deep drilling on Midway Atoll: U.S. Geol. Survey Prof. Paper 680-A, p. 1-22.
——— ——— J. W. Wells, and K. O. Emery, 1950, Organic growth and sedimentation on an atoll (Bikini): Jour. Geology, v. 58, no. 4, p. 410-425.
Lalou, C., J. Labeyrie, and G. Delibrias, 1970, Datation des calcaires coralliens de l'atoll de Mururoa (archipel des Tuamotu) de l'epoque actuelle jusqu'a, 500,000 ans: Acad. Sci. Comptes Rendus, ser. D, v. 263, no. 25, p. 1946-1949.

Land, L. S., and T. F. Goreau, 1970, Submarine lithification of Jamaican reefs: Jour. Sed. Petrology, v. 40, no. 1, p. 457-462.
——— F. T. Mackenzie, and S. J. Gould, 1967, Pleistocene history of Bermuda: Geol. Soc. America Bull., v. 78, no. 8, p. 993-1006.
Layer, D. B., *et al.*, 1949, Leduc oil field, Alberta, a Devonian coral-reef discovery: Am. Assoc. Petroleum Geologists Bull., v. 33, no. 4, p. 572-602.
Lecompte, M., 1938, Quelques types de "recifs" Siluriens et Devoniens del'Amerique du Nord; essai de comparaison avec les recifs actuels: Mus. Royal Histoire Nat. Belgique Bull., v. 14, no. 39, p. 1-51.
——— 1958, Les recifs paleozoiques en Belgique: Geol. Rundschau, v. 47, no. 1, p. 384-401.
——— 1959, Certain data on the genesis and ecologic character of Frasnian reefs of the Ardennes: Internat. Geology Rev., v. 1, no. 7, p. 1-14.
——— 1970, Der Riffe im Devon der Ardennen und ihre Bildungsbedingungen: Geologica et Palaeontologica, v. 4, p. 25-65.
Lees, A., 1964, The structure and origin of the Waulsortian (Lower Carboniferous) "reefs" of west central Eire: Royal Soc. London Philos. Trans., ser. B, v. 247, p. 483-531.
Lewis, J. B., 1960, The coral reefs and coral communities of Barbados, West Indies: Canadian Jour. Zoology, v. 38, p. 1133-1145.
——— 1965, A preliminary description of some marine benthic communities from Barbados, West Indies: Canadian Jour. Zoology, v. 43, no. 5, p. 1049-1074.
Lewis, M. S., 1969, Sedimentary environments and unconsolidated carbonate sediments of the fringing coral reefs of Mahe, Seychelles: Marine Geology, v. 7, no. 2, p. 95-127.
Link, T. A., 1950, Theory of transgressive and regressive reef (bioherm) development and origin of oil: Am. Assoc. Petroleum Geologists Bull., v. 34, no. 2, p. 263-294.
Logan, B. W., 1969, Coral reefs and banks, Yucatan shelf, Mexico, pt. 2, *in* Carbonate sediments and reefs, Yucatan shelf, Mexico: Am. Assoc. Petroleum Geologists Mem. 11, p. 129-198.
——— R. Rezak, and R. N. Ginsburg, 1964, Classification and environmental significance of algal stromatolites: Jour. Geology, v. 72, p. 68-83.
Lowenstam, H. A., 1948, Marine pool, Madison County, Illinois, Silurian reef producer, *in* Structure of typical American oil fields: Am. Assoc. Petroleum Geologists, p. 153-188.
——— 1950, Niagaran reefs of the Great Lakes area: Jour. Geology, v. 58, no. 4, p. 430-487.
——— 1957, Niagaran reefs in the Great Lakes area, Chap. 10 *of* H. S. Ladd, ed., Paleoecology: Geol. Soc. America, Mem. 67, v. 2, p. 215-248.
Lum, D., and H. T. Stearns, 1970, Pleistocene stratigraphy and eustatic history based on core at Waimanalo, Oahu, Hawaii: Geol. Soc. America Bull., v. 81, no. 1, p. 1-16.
Macintyre, I. G., 1972, Submerged reefs of eastern Caribbean: Am. Assoc. Petroleum Geologists Bull., v. 56, no. 4, p. 720-738.
MacNeil, F. S., 1954a, Organic reefs and banks and associated detrital sediments: Am. Jour. Sci., v. 252, no. 7, p. 385-401.
——— 1954b, The shape of atolls—an inheritance from subaerial erosion forms: Am. Jour. Sci., v. 252, no. 7, p. 402-427.
Maksimova, S. V., 1972, Coral reefs in the Arctic and their paleogeographical interpretation: Internat. Geology Rev., v. 14, no. 7, p. 764-769.
Manton, S. M., and T. A. Stephenson, 1935, Ecological surveys of coral reefs: British Mus. (Nat. History) Sci. Repts. Great Barrier Reef Expedition 1928-1929, v. 3, no. 10, p. 273-312.
Maxwell, W. G. H., 1968, Atlas of the Great Barrier Reef: Amsterdam, Elsevier, 258 p.
——— 1970, Deltaic patterns in reefs: Deep Sea Research, v. 17, no. 6, p. 1005-1018.
Mayor, A. G., 1924, Structure and ecology of Samoan reefs: Paper Carnegie Inst. Washington Dept. Marine Biology, v. 19, p. 1-25.
Mergner, H., 1971, Structure, ecology, and zonation of Red Sea reefs (in comparison with south Indian and Jamaican reefs), *in* Regional variation in Indian Ocean coral reefs: Zool. Soc. London Symp. 28, p. 141-161.
Mesolella, K. J., R. K. Matthews, W. S. Broeker, and D. L. Thurber, 1969, The astronomical theory of climatic change: Barbados data: Jour. Geology., v. 77, no. 3, p. 250-274.
——— H. A. Sealy, and R. K. Matthews, 1970, Facies geometries within Pleistocene reefs of Barbados, West Indies: Am. Assoc. Petroleum Geologists Bull., v. 54, no. 10, p. 1899-1917.
Miller, J., and R. F. Grayson, 1972, Origin and structure of the lower Visean "reef" limestones near Clitheroe, Lancashire: Yorkshire Geol. Soc. Proc., v. 38, no. 4, p. 607-638.
Milliman, J. D., 1965, An annotated bibliography of recent papers on corals and coral reefs: Atoll Research Bull., v. 111, p. 1-58.
——— 1967, The geomorphology and history of Hogsty Reef, a Bahamian atoll: Bull. Marine Science, v. 17, p. 519-543.
Molinier, R., and J. Picard, 1952, Recherches sur les herbiers de phanerogames marines du littoral Mediterranean Francais: Inst. Oceanog. Annales, v. 27, no. 3, p. 157-234.
Montaggioni, L., 1970a, Essai de chronologie relative des episodes recifaux a l'ile de le Reunion (Ocean Indien); leur incidence sur la morphologie recifale actuelle: Acad. Sci. Comptes Rendus, ser. D, v. 270, no. 15, p. 1869-1871.
——— 1970b, Essai de reconstitution paleogeographique de l'ile Rodrique (Archipel des Mascareignes, Ocean Indien): Acad. Sci. Comptes Rendus, ser. D, v. 271, p. 1741-1744.
Morton, J. E., and D. Challis, 1969, The biomorphology of Solomon Island shores, with a discussion of zoning patterns and ecological terminology: Royal Soc. London Philos. Trans., ser. B, v. 255, p. 459-516.
Nelson, H. F., C. W. Brown, and J. H. Brineman, 1962, Skeletal limestone classifications, *in* Classification of carbonate rocks: Am. Assoc. Petroleum Geologists Mem. 1, p. 224-252.
Newell, N. D., 1971, An outline history of tropical organic reefs: Am. Mus. Novitates, v. 2465, p. 1-37.
——— and J. K. Rigby, 1957, Geological studies on the Great Bahama Bank, *in* Regional aspects of carbonate deposition: Soc. Econ. Paleontologists and Mineralogists Spec. Pub. 5, p. 15-79.
——— ——— A. G. Fischer, A. J. Whiteman, J. E. Hickox, and J. S. Bradley, 1953, The Permian reef complex of the Guadalupe Mountains region, Texas and New Mexico: a study in paleoecology: San Francisco, W. H. Freeman, 236 p.
Norris, R. M., 1953, Buried oyster reefs in some Texas bays: Jour. Paleontology, v. 27, no. 4, p. 569-576.
Nugent, L. E., Jr., 1946, Coral reefs in the Gilbert, Marshall, and Caroline Islands: Geol. Soc. America Bull., v. 57, no. 8, p. 735-779.
Odum, H. T., and E. P. Odum, 1955, Trophic structure and productivity of a windward coral reef community on Eniwetok Atoll: Ecol. Monogr., v. 25, p. 291-320.
Picard, J., 1967, Essai de classement des grands types de peuplements marine benthiques tropicaux d'apres les observations effectuees dans les parages de Tulear (sud-ouest de Madagascar): Station Marine Endoume Recueil Travaux, v. 4 (fasc. hors serie, supp. 6), p. 3-24.
Pichon, M., 1971, Comparative study of the main features of some coral reefs of Madagascar, La Reunion, and Mauritius, *in* Regional variation in Indian Ocean coral reefs: Zool. Soc. London Symp. 28, p. 185-216.
Pray, L. C., 1958, Fenestrate bryozoan core facies, Mississippian bioherms, southwestern United States: Jour. Sed. Petrology, v. 28, no. 3, p. 261-273.
Pugh, W. E., ed., 1950, Bibliography of organic reefs, bioherms, and biostromes: Tulsa, Oklahoma, Seismograph Service Corp., 139 p.
Richards, H. C., and D. Hills, 1942, Great Barrier Reef bores,

1926 and 1937: descriptions, analyses, and interpretations: Great Barrier Reef Comm. Rept., v. 5, p. 1-111.

Riding, R., and D. F. Toomey, 1972, The sedimentological role of *Epiphyton* and *Renalcis* in Lower Ordovician mounds, southern Oklahoma: Jour. Paleontology, v. 46, no. 4, p. 509-519.

Roll, A., 1934, Form, Bau und Entstehung der Schwammostotzen in suddeutschen Malm: Palaont. Zeitschr., v. 16, p. 197-246.

Rosen, B. R., 1971, Principal features of reef-coral ecology in shallow-water environments of Mahe, Seychelles, *in* Regional variation in Indian Ocean coral reefs: Zool. Soc. London Symp. 28, p. 163-183.

Sanders, J. E., and G. M. Friedman, 1967, Origin and occurrence of limestones, *in* Carbonate rocks: Amsterdam, Elsevier, Developments in sedimentology, v. 9A, p. 169-265.

Schlanger, S. O., 1963, Subsurface geology of Eniwetok Atoll: U.S. Geol. Survey Prof. Paper 260-BB, p. 991-1066.

Scoffin, T. P., 1970, The trapping and binding of subtidal carbonate sediments by marine vegetation in Bimini Lagoon, Bahamas: Jour. Sed. Petrology, v. 40, no. 1, p. 249-273.

——— 1971, The conditions of growth of the Wenlock reefs of Shropshire (England): Sedimentology, v. 17, no. 3, p. 173-219.

Sestini, J., 1965, Cenozoic stratigraphy and depositional history, Red Sea coast, Sudan: Am. Assoc. Petroleum Geologists Bull., v. 49, no. 9, p. 1453-1472.

Sheer, G., 1971, Coral reefs and coral genera in the Red Sea and Indian Ocean, *in* Regional variation in Indian Ocean coral reefs: Zool. Soc. London Symp. 28, p. 329-367.

Shinn, E. A., 1963, Spur and groove formation on the Florida reef tract: Jour. Sed. Petrology, v. 33, no. 2, p. 291-303.

Smith, D. B., and E. A. Francis, 1967, Geology of the country between Durham and West Hartlepool: Great Britain Geol. Survey Mem. 27, p. 1-354.

Stanley, S. M., 1966, Paleoecology and diagenesis of Key Largo Limestone, Florida: Am. Assoc. Petroleum Geologists Bull., v. 50, no. 9, p. 1927-1947.

——— 1968, Post-Paleozoic adaptive radiation of infaunal bivalve molluscs—a consequence of mantle fusion and siphon formation: Jour. Paleontology, v. 42, no. 1, p. 214-229.

Stanley, D. J., and D. J. P. Swift, 1968, Bermuda's reef-front platform; bathymetry and significance: Marine Geology, v. 6, no. 6, p. 479-500.

Stanton, R. J., Jr., 1967, Factors controlling shape and internal facies distribution of organic carbonate buildups: Am. Assoc. Petroleum Geologists Bull., v. 51, no. 12, p. 2462-2467.

Stenzel, H. B., 1971, Oysters, *in* Treatise on invertebrate paleontology (N) Mollusca, v. 3: Geol. Soc. America, p. N953-N1224.

Stephenson, T. A., A. Stephenson, G. Tandy, and M. Spender, 1931, The structure and ecology of Low Isles and other reefs: British Mus. (Nat. History) Sci. Rept. Great Barrier Reef Expedition 1928-1929, v. 3, no. 2, p. 167-217.

Stetson, T. R., D. F. Squires, and R. M. Pratt, 1962, Coral banks occurring in deep water on the Blake Plateau: Am. Mus. Novitates, no. 2114, 39 p.

Stoddart, D. R., 1969a, Ecology and morphology of recent coral reefs: Biol. Rev., v. 44, no. 4, p. 433-498.

——— 1969b, Geomorphology of the Solomon Islands coral reefs: Royal Soc. London Philos. Trans., ser. B, v. 255, p. 355-382.

Storr, J. F., 1964, Ecology and oceanography of the coral-reef tract, Abaco Island, Bahamas: Geol. Soc. America Spec. Paper 79, 98 p.

Taft, W. H., F. Arrington, A. Haimovitz, C. MacDonald, and C. Woolheater, 1968, Lithification of modern marine carbonate sediments at Yellow Bank, Bahamas: Bull. Marine Sci., v. 18, no. 4, p. 762-828.

Taylor, J. D., 1968, Coral reef and associated invertebrate communities (mainly molluscan) around Mahe, Seychelles: Royal Soc. London Philos. Trans., ser. B, v. 254, p. 129-206.

Taylor, P. W., 1951, The Plymouth limestones: Royal Geol. Soc. Cornwall Trans., v. 18, no. 2, p. 146-160.

Teichert, C., 1958, Cold- and deep-water coral banks: Am. Assoc. Petroleum Geologists Bull., v. 42, no. 5, p. 1064-1082.

——— and R. W. Fairbridge, 1948, Some coral reefs of the Sahul shelf: Geog. Rev., v. 38, no. 2, p. 222-249.

Terry, C. E., and J. J. Williams, 1969, The Idris "A" bioherm and oilfield, Sirte basin, Libya; its commercial development, regional Palaeocene geologic setting and stratigraphy, *in* Exploration for petroleum in Europe and north Africa: London, Inst. Petroleum, p. 31-48.

Textoris, D. A., and A. V. Carozzi, 1964, Petrography and evolution of Niagaran (Silurian) reefs, Indiana: Am. Assoc. Petroleum Geologists Bull., v. 48, no. 4, p. 397-426.

Thomassin, B. A., 1969, Peuplements de deux biotopes de sables coralliens sur le Grande Recif de Tulear (S.W. de Madagascar): Station Marine Endoume Recueil Travaux (fasc. hors serie, supp. 9), p. 59-133.

Thurber, D. L., W. S. Broeker, H. A. Potratz, and R. L. Blanchard, 1965, Uranium series ages of Pacific atoll coral: Science, v. 149, p. 55-58.

Tracey, J. I., H. S. Ladd, and J. E. Hoffmeister, 1948, Reefs of Bikini, Marshall Islands: Geol. Soc. America Bull., v. 59, no. 9, p. 861-878.

Tsien, H. H., 1971, The Middle and Upper Devonian reef-complexes of Belgium: Petroleum Geology Taiwan, no. 8, p. 119-173.

Umbgrove, J. H. F., 1939, Madreporaria from the Bay of Batavia: Zoolog. Med., v. 22, p. 1-64.

——— 1947, Coral reefs of the East Indies: Geol. Soc. America Bull., v. 58, no. 8, p. 729-778.

Vacelet, J., and P. Vasseur, 1971, Eponges des recifs coralliens de Tulear (Madagascar): Station Marine d'Endoume Tethys, supp. 1, p. 51-126.

Vaughan, T. W., 1919, Conditions under which the West Indian, Central American, and Floridian coral reefs have formed, and their bearing on theories of coral reef formation, *in* Contributions to the geology and paleontology of the Canal Zone, Panama, and geologically related areas in Central America and the West Indies: U.S. Natl. Mus. Bull. 103, p. 238-332.

Walker, K. R., 1972, Community ecology of the Middle Ordovician Black River Group of New York State: Geol. Soc. America Bull., v. 83, no. 8, p. 2499-2524.

Wells, J. W., 1957, Coral reefs, Chap. 20, *of* Ecology: Geol. Soc. America Mem. 67, v. 1, p. 609-631.

Weydert, P., 1969, Les variations recentes du niveau marin et leurs influences sur la morphologie recifale dans la baie de Tulear (sud-ouest de Madagascar): Acad. Sci. Comptes Rendus, ser. D, v. 268, p. 482-484.

Wilson, V., 1949, The lower Corallian rocks of the Yorkshire coast and Hackness Hills: Geologists Assoc. Proc., v. 60, pt. 4, p. 235-271.

Wilson, W. B., 1950, Reef definition: Am. Assoc. Petroleum Geologists Bull., v. 34, no. 2, p. 181.

Worth, R. N., 1888, The limestone of the Plymouth district: Devonshire Assoc. Trans., v. 20, p. 410-423.

Yabe, H., and T. Sugiyama, 1932, Reef corals found in the Japanese seas: Tohoku Univ. Sci. Repts., ser. 2, v. 15, pt. 2, p. 143-168.

Zankl, H., 1969, Der Hohe Goll; aufbau und Lebensbild eines Dachsteinkalk-Riffes in der Obertrias der nordlichen Kalkalpen: Senckenbergische Naturf. Gesell., Abh., no. 519, 96 p.

——— 1971, Upper Triassic carbonate facies in the northern Limestone Alps, *in* Sedimentology of parts of Central Europe: 8th Internat. Sedimentation Cong. Guidebook, Heidelberg, p. 147-185.